Nuclear Science and Technology

Nuclear Science and Technology

Edited by
Christine Avory

Nuclear Science and Technology
Edited by Christine Avory
ISBN: 978-1-63549-685-7 (Hardback)

Published by Larsen and Keller Education,
5 Penn Plaza,
19th Floor,
New York, NY 10001, USA

Cataloging-in-Publication Data

Nuclear science and technology / edited by Christine Avory.
p. cm.
Includes bibliographical references and index.
ISBN 978-1-63549-685-7
1. Nuclear energy. 2. Nuclear engineering. 3. Nuclear energy--Technological innovations.
4. Nuclear physics. I. Avory, Christine.
TK9145 .N83 2018
621.48--dc23

For more information regarding Larsen and Keller Education and its products, please visit the publisher's website www.larsen-keller.com

Table of Contents

Preface

As an important part of nuclear science, nuclear reactor technology refer to the device which has the power to start, sustain and control a nuclear chain reaction. These devices are built on the technology of nuclear fission, reactivity control, heat generation, electrical power, etc. They are also used for electricity generation at nuclear power plants. This book aims to equip students in the field of nuclear science and technology. The topics covered in it are of utmost significance and are bound to provide thorough knowledge to the readers about this area. Different approaches, evaluations and methodologies on nuclear technology have been included in this textbook. Those in search of information to further their knowledge will be greatly assisted by it.

A short introduction to every chapter is written below to provide an overview of the content of the book:

Chapter 1 - Nuclear reactors are used for the generation of electricity and in the propulsion of ships. Some of the key components of nuclear power plants are nuclear reactor core, neutron poison, coolant, control rods, startup neutron source and boiler feedwater pump. This is an introductory chapter which will introduce briefly all the significant aspects of nuclear science and technology; **Chapter 2 -** The fuel used in nuclear power stations to produce heat required for turbines is known as nuclear fuel. Nuclear fuel cycle is the process of refining and purifying nuclear fuel. The chapter strategically encompasses and incorporates the major components and key concepts of nuclear science and technology, providing a complete understanding; **Chapter 3 -** The design constraints related to nuclear reactors are the maximum temperature of clad, linear (heat/power) rating, coolant velocity, etc. Corrosion is also an important factor to be kept in mind while designing a nuclear reactor. This chapter elucidates the crucial theories and principles of nuclear science and technology; **Chapter 4 -** Boiling water reactor is used for the production of electrical power. It is a common type of nuclear reactor that is used in generating electricity. The other type of thermal reactor discussed in this chapter is gas-cooled reactor. The aspects elucidated in this chapter are of vital importance, and provide a better understanding of nuclear science and technology; **Chapter 5 -** Nuclear fast reactor is a type of nuclear reactor. Instead of using neutron moderator, nuclear fast reactors use fast neutrons. Water too, is not used as a coolant in these types of reactors. In order to completely understand nuclear science and technology, it is necessary to understand the processes related to it. The following chapter elucidates the varied processes and mechanisms associated with this area of study; **Chapter 6 -** Heat transfer is the interchange of thermal energy between two physical systems. It can be classified into numerous mechanisms like thermal convection, thermal conduction and thermal radiation. Nuclear fission and fusion generate great amounts of heat. The chapter closely examines the key concepts of heat flow process to provide an extensive understanding of the subject; **Chapter 7 -** In fast reactors, the neutron energy is released in a wide spectrum. Unmoderated fast neutrons are used in these reactors since they have a better fission/capture ratio, making fast reactors breed more fissile fuel than it consumes. Nuclear science and technology is best understood in confluence with the major topics listed in the following chapter.

Finally, I would like to thank my fellow scholars who gave constructive feedback and my family members who supported me at every step.

Editor

1

Nuclear Reactor: An Introduction

Nuclear reactors are used for the generation of electricity and in the propulsion of ships. Some of the key components of nuclear power plants are nuclear reactor core, neutron poison, coolant, control rods, startup neutron source and boiler feedwater pump. This is an introductory chapter which will introduce briefly all the significant aspects of nuclear science and technology.

Nucleus

We are aware that protons and neutrons comprise nucleus of an atom. Nucleus is positive, due to the presence of positively charged protons and the neutrons (no charge). Few elements have same atomic number 'Z' (same number of electrons) but possess different mass number 'A' (total sum of electrons and neutrons). Such elements are called Isotopes, the lightest of them being ${}^{2}_{1}D$ (Deuterium) and ${}^{3}_{1}T$ (Tritium) being the isotopes of ${}^{1}_{1}H$ (Hydrogen) as shown in Figure.

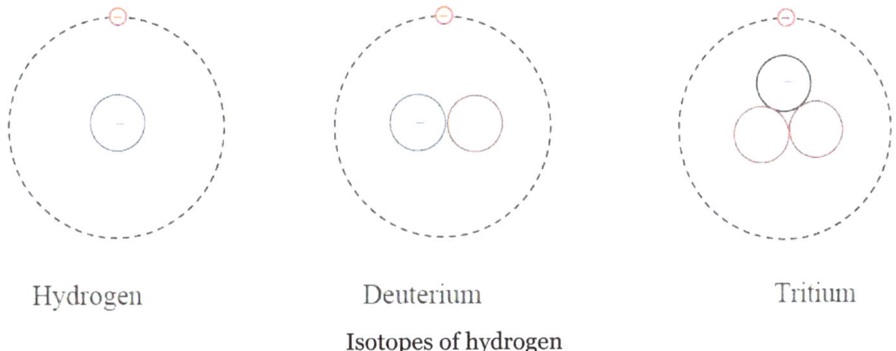

Isotopes of hydrogen

The difference between the atomic mass (A) and the atomic number (Z) is the neutron number (N). Hence, the isotopes of same element possess different neutron numbers (N). For the purpose of distinguishing isotopes, 'A' may be identified as the total number of nucleons, which are particles heavier than electron (proton and neutron are heavier than electron).

The size of a nucleus is much lesser than the size of an atom. Nucleus is denser than the atom, due to its smaller volume and larger mass of protons. The radius of a nucleus is related to the number of nucleons (A) as follows:

$$R(cm) = 1.4 \times 10^{-13} A^{1/3}$$

The mass of proton, neutron, electron and atom are expressed in amu (atomic mass unit). One amu is equal to 1.66 x 10^{-24} g. Table gives the mass of a neutron, proton, and an electron.

Table: Masses of neutron, proton and electron

Particle	Mass (amu)
Neutron	1.008665
Proton	1.007276
Electron	0.000549

Binding Energy of a Nucleus (BE)

As the nucleus contains protons (all of which are positively charged) and neutrons (no charge), one would expect electrostatic repulsion between protons to be high. Electrostatic repulsion between two charged particles varies inversely with the square of distance between them. Hence, as two protons come close to each other, one would expect electrostatic repulsion to make them move apart. However, the existence of nucleus testifies the fact that there exists a short- range attractive force stronger than the electrostatic repulsion. This short-range force is the nuclear force and is responsible for binding nucleons into a compact nucleus. A potential energy of binding (called binding energy) is associated with this force and hence energy equivalent to this must be supplied to disrupt a nucleus to component nucleons.

The binding energy, denoted as BE, of a nucleus may be calculated using Einstein's mass-energy relation as follows:

Let m_n be the mass of a neutron, m_H be the sum of mass of a proton and an electron, then the total mass of all constituent particles (m_{CP}) of an atom with atomic number (Z) and neutron number (N) is given by

$$m_{CP} = Z(m_H) + N(m_n) \qquad (1)$$

If mass of the atom is M, then the difference between the total mass of all constituent particles and that of the atom is the mass defect (m_d), given by

$$m_d = m_{CP} - M = Z(m_H) + N(m_n) - M \qquad (2)$$

Taking 'm_H' and 'm_n' as 1.007825 and 1.008665 amu, we get the mass defect as

$$m_d = Z(1.007825) + N(1.008665) - M \qquad (3)$$

The unit of m_d is also amu.

Using Einstein's relationship, one may relate energy and mass as $E = mc^2$ (4)

Taking 'm' as 1 amu, E = 1.494×10^{-10} J = 931 MeV (1 eV = 1.6×10^{-19} J)

For an atom with a mass defect 'm_d', binding energy = $m_d c^2$ (5)

From the tabulated values of Z, N and M for an atom the binding energy may be calculated using the equations (2) and (5)

Average Nuclear Binding Energies

Average nuclear binding energy = BE/A = Binding energy of a nucleus/number of nucleons

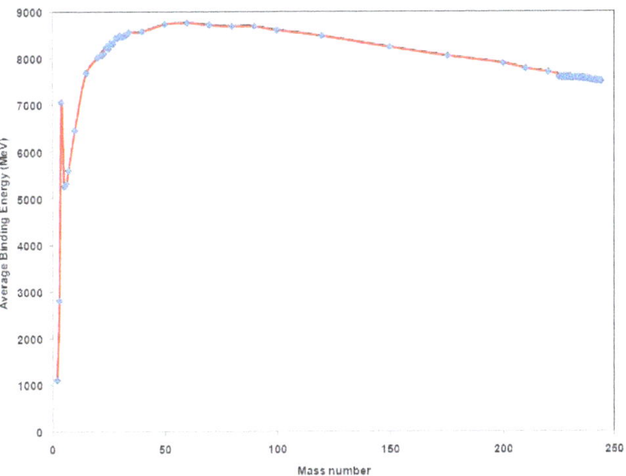

Average binding energy as a function of mass number

The elements with higher binding energy per nucleon are difficult to break up. Iron with a mass number of 56 has the highest binding energy per nucleon (average binding energy). Among the light elements Helium with a mass number of 4 has the highest average binding energy.

For data on atomic masses and average binding energies of nucleus of various isotopes, one may refer to "LBNL Isotopes Project Nuclear Data Dissemination Home Page"

Example -1: Determine the binding energy of the nucleus in $^{235}{}_{92}U$. The mass of ^{235}U atom is 235.0439231 amu.

Recalling Eq. (2), m_d = Z(1.007825)+N(1.008665)-M

Z= 92; A = 235; N = A-Z = 143, M = 235.0439231 amu

Substituting above in Equation (2) yields a mass defect (m_d) of 1.9151 amu.

Recalling that 1 amu corresponds to 931 MeV, the binding energy of U-235 is 1.915*931 ~ 1793 MeV

Example – 2: Determine the average binding energy of the nucleus in $^{239}{}_{94}Pu$. The mass of ^{239}Pu is 239.052163 amu.

Recalling Equation (2), m_d = Z(1.007825)+N(1.008665)-M

Z= 94; A = 239; N = A-Z = 145, M = 239.052163 amu

Substituting above in Eq. (2) yields a mass defect (m_d) of 1.9398 amu.

Recalling that 1 amu corresponds to 931 MeV, the binding energy of U-235 is 1.9398*931 ~ 1805.965 MeV

Average binding energy = binding energy/number of nucleons = 1805.965/239 = 7.556 MeV

Binding Energy of a Neutron

Similar to the calculation of binding energy of a nucleus, binding energy associated with addition of a neutron (B_n) of the following type of neutron reaction can be calculated.

$$^{15}_{8}O + ^{1}_{0}n \rightarrow\rightarrow ^{16}_{8}O$$

$$m_{dnu} = M_1 + m_n - M_2 \quad (6)$$

In Equation (5) 'M_1' and 'M_2' are the mass of atom before ($^{15}_{8}O$) and after neutron $^{16}_{8}O$ absorption. 'm_{dnu}' is the mass defect of a neutron.

Example - 3: Determine the binding energy of the neutron added during the following reaction:

$$^{16}_{8}O + ^{1}_{0}n \rightarrow\rightarrow ^{17}_{8}O$$

Recall Eq. (5) $m_{dnu} = M_1 + mn - M_2$

Mass of $^{16}_{8}O$ = M_1 = 15.9949146; Mass of $^{17}_{8}O$ = M_2 =16.9991317;

m_n=1.008665

Substituting above in Eq. (5) yields,

m_{dnu}= 0.004448 amu

Binding energy = $m_{dnu}c^2$ = 4.141 MeV

Neutron Classification

Depending on the kinetic energy of a neutron, it may be classified into number of categories chief among which are thermal and fast neutrons.

Neutrons with kinetic energies of about 0.025 eV are called thermal electrons, while those with kinetic energies around 1 MeV.

Recalling the relationship between speed and kinetic energy, speed of a neutron (u_n) corresponding to a kinetic energy (E_n) can be calculated using the following equation:

$$0.5 m_n u_n^2 = E_n \quad (7)$$

From Eq. (7) At 20 °C, the speeds of thermal and fast neutrons are 2.2 km/s and 14000 km/s respectively.

Nuclear Reactions

These are the reactions involving the participation of atomic nuclei. Some of these reactions are spontaneous (Eg. Radioactivity) while few others are result of bombardment of nuclei with an energetic particle or radiation.

There are certain similarities and differences between nuclear and chemical reactions. The similarity lies in the fact that both these type of reactions obey the conservation laws: conservation of mass, energy, particles and the charge.

The difference lies in the amount of energy released. Nuclear reactions are highly energetic compared to chemical reactions. For example, fission of an atom of Uranium-235 releases about 210 MeV of energy, while energy released due to the formation of one molecule of CO_2 from the combustion of carbon is about 4.8 eV.

The types of nuclear reactions that are of importance in nuclear reactors are:

(i) Elastic scattering

(ii) Inelastic scattering

(iii) Neutron capture

(iv) Fission

Elastic Scattering

This is a type of neutron (nuclear) reaction involving elastic collision of a neutron with a nucleus. The structure or mass of nucleus does not change as a result of this reaction. However, the speed of the neutron and its direction changes. As the name 'elastic' scattering suggests there is no change in kinetic energy of the system during this event.

In nutshell, Elastic scattering is change in speed and direction of a neutron (scattering) without any change in total kinetic energy of the system (elastic), though the neutron loses a part of its kinetic energy and slows down.

The elastic scattering utilized in the nuclear reactors is the scattering of neutron by reaction with light nuclei (hydrogen, deuterium, carbon) resulting in a large loss of kinetic energy of neutron, without any loss in the number of neutrons. This is called 'moderation' and the substances containing light nuclei (water, heavy water, graphite) are called 'moderators'.

Inelastic Scattering

This is a type of neutron (nuclear) reaction involving inelastic collision of a neutron with a nucleus. The nucleus reaches an excited state upon neutron irradiation, which decays rapidly to the ground state. This is accompanied by the emission of gamma radiation.

$$n + {}^{238}\mathrm{U} \rightarrow n + {}^{238}\mathrm{U}^* \rightarrow n + {}^{238}\mathrm{U} + \gamma$$

'γ' represents gamma radiation and ^{238}U is the target nucleus

For inelastic scattering to occur, the energy of the neutron incident on a nucleus must necessarily be greater than the excitation energy of the lowest excited state of the nucleus. The initial slowing down of neutron in a reactor system is due to inelastic scattering.

Neutron Capture

Different from elastic and inelastic scattering where there is no change in the atomic mass of target nucleus, neutron capture results in the increase in mass of target nucleus which reaches the excited state. Let us consider the following example

$$n + {}^{238}\text{U} \rightarrow {}^{239}\text{U}^*$$

The excited nucleus then decays rapidly (within 10^{-13} seconds) by emission of one or gamma radiation as follows:

$$^{239}\text{U}^* \rightarrow {}^{239}\text{U} + \gamma$$

The importance of neutron capture in a nuclear reactor is as follows:

(i) Neutrons are captured by control rods which otherwise might initiate fission

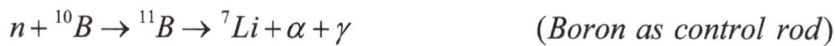

$$n + {}^{113}Cd \rightarrow {}^{113}Cd + \gamma \qquad (Cadmium\ as\ control\ rod)$$

$$n + {}^{10}B \rightarrow {}^{11}B \rightarrow {}^{7}Li + \alpha + \gamma \qquad (Boron\ as\ control\ rod)$$

(ii) Transformation of one nucleus to the other is due to neutron capture, enabling production of isotopes of interest

$$n + {}^{59}Co \rightarrow {}^{60}Co + \gamma \qquad (Radioactive\ cobalt\ for\ medicinal\ applications)$$

$$n + {}^{238}\text{U} \rightarrow {}^{239}\text{U} \rightarrow {}^{239}Np + {}^{0}_{-1}e$$

$$^{239}Np \rightarrow {}^{239}Pu + {}^{0}_{-1}e$$

The half-lives of ^{239}U and ^{239}Np are 23.5 min and 2.355 days respectively, shorter than the time involved in the production of fissile material ^{239}Pu. The decay of ^{239}Pu, an isotope with a high half-life of about 2.41x104 years, is not appreciable in the reactor. The isotope that undergoes nuclear transmutation to produce a fissile isotope is called fertile isotope. In the above example 238U is the fertile isotope, which upon nuclear transmutation produces ^{239}Pu.

Nuclear Reactor

A nuclear reactor, formerly known as an atomic pile, is a device used to initiate and control a sustained nuclear chain reaction. Nuclear reactors are used at nuclear power plants for electricity generation and in propulsion of ships. Heat from nuclear fission is passed to a working fluid (water or gas), which runs through steam turbines. These either drive a ship's propellers or turn electrical generators. Nuclear generated steam in principle can be used for industrial process heat or for district heating. Some reactors are used to produce isotopes for medical and industrial use, or for production of weapons-grade plutonium. Some are run only for research. As of April 2014, the IAEA reports there are 435 nuclear power reactors in operation, in 31 countries around the world.

Core of CROCUS, a small nuclear reactor used for research at the EPFL in Switzerland

Mechanism

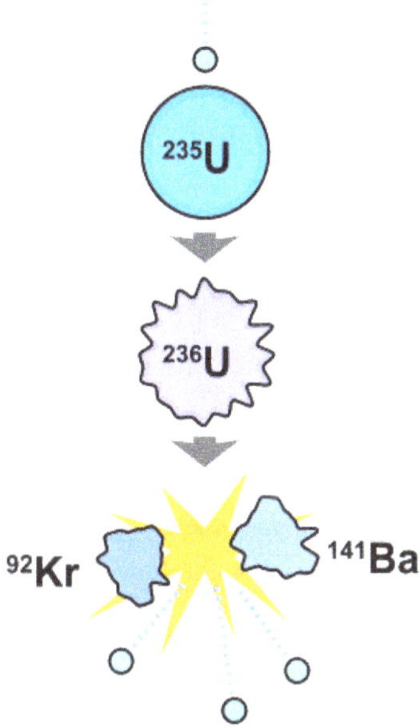

An induced nuclear fission event. A neutron is absorbed by the nucleus of a uranium-235 atom, which in turn splits into fast-moving lighter elements (fission products) and free neutrons. Though both reactors and nuclear weapons rely on nuclear chain-reactions, the rate of reactions in a reactor occurs much more slowly than in a bomb.

Just as conventional power-stations generate electricity by harnessing the thermal energy released from burning fossil fuels, nuclear reactors convert the energy released by controlled nuclear fission into thermal energy for further conversion to mechanical or electrical forms.

Fission

When a large fissile atomic nucleus such as uranium-235 or plutonium-239 absorbs a neutron, it may undergo nuclear fission. The heavy nucleus splits into two or more lighter nuclei, (the fission products), releasing kinetic energy, gamma radiation, and free neutrons. A portion of these neutrons may later be absorbed by other fissile atoms and trigger further fission events, which release more neutrons, and so on. This is known as a nuclear chain reaction.

To control such a nuclear chain reaction, neutron poisons and neutron moderators can change the portion of neutrons that will go on to cause more fission. Nuclear reactors generally have automatic and manual systems to shut the fission reaction down if monitoring detects unsafe conditions.

Commonly-used moderators include regular (light) water (in 74.8% of the world's reactors), solid graphite (20% of reactors) and heavy water (5% of reactors). Some experimental types of reactor have used beryllium, and hydrocarbons have been suggested as another possibility.

Heat generation

The reactor core generates heat in a number of ways:

- The kinetic energy of fission products is converted to thermal energy when these nuclei collide with nearby atoms.
- The reactor absorbs some of the gamma rays produced during fission and converts their energy into heat.
- Heat is produced by the radioactive decay of fission products and materials that have been activated by neutron absorption. This decay heat-source will remain for some time even after the reactor is shut down.

A kilogram of uranium-235 (U-235) converted via nuclear processes releases approximately three million times more energy than a kilogram of coal burned conventionally (7.2×10^{13} joules per kilogram of uranium-235 versus 2.4×10^{7} joules per kilogram of coal).

Cooling

A nuclear reactor coolant — usually water but sometimes a gas or a liquid metal (like liquid sodium) or molten salt — is circulated past the reactor core to absorb the heat that it generates. The heat is carried away from the reactor and is then used to generate steam. Most reactor systems employ a cooling system that is physically separated from the water that will be boiled to produce pressurized steam for the turbines, like the pressurized water reactor. However, in some reactors the water for the steam turbines is boiled directly by the reactor core; for example the boiling water reactor.

Reactivity Control

The power output of the reactor is adjusted by controlling how many neutrons are able to create more fissions.

Control rods that are made of a neutron poison are used to absorb neutrons. Absorbing more neutrons in a control rod means that there are fewer neutrons available to cause fission, so pushing the control rod deeper into the reactor will reduce its power output, and extracting the control rod will increase it.

At the first level of control in all nuclear reactors, a process of delayed neutron emission by a number of neutron-rich fission isotopes is an important physical process. These delayed neutrons account for about 0.65% of the total neutrons produced in fission, with the remainder (termed "prompt neutrons") released immediately upon fission. The fission products which produce delayed neutrons have half lives for their decay by neutron emission that range from milliseconds to as long as several minutes, and so considerable time is required to determine exactly when a reactor reaches the critical point. Keeping the reactor in the zone of chain-reactivity where delayed neutrons are *necessary* to achieve a critical mass state allows mechanical devices or human operators to control a chain reaction in "real time"; otherwise the time between achievement of criticality and nuclear meltdown as a result of an exponential power surge from the normal nuclear chain reaction, would be too short to allow for intervention. This last stage, where delayed neutrons are no longer required to maintain criticality, is known as the prompt critical point. There is a scale for describing criticality in numerical form, in which bare criticality is known as *zero dollars* and the prompt critical point is *one dollar*, and other points in the process interpolated in cents.

In some reactors, the coolant also acts as a neutron moderator. A moderator increases the power of the reactor by causing the fast neutrons that are released from fission to lose energy and become thermal neutrons. Thermal neutrons are more likely than fast neutrons to cause fission. If the coolant is a moderator, then temperature changes can affect the density of the coolant/moderator and therefore change power output. A higher temperature coolant would be less dense, and therefore a less effective moderator.

In other reactors the coolant acts as a poison by absorbing neutrons in the same way that the control rods do. In these reactors power output can be increased by heating the coolant, which makes it a less dense poison. Nuclear reactors generally have automatic and manual systems to scram the reactor in an emergency shut down. These systems insert large amounts of poison (often boron in the form of boric acid) into the reactor to shut the fission reaction down if unsafe conditions are detected or anticipated.

Most types of reactors are sensitive to a process variously known as xenon poisoning, or the iodine pit. The common fission product Xenon-135 produced in the fission process acts as a neutron poison that absorbs neutrons and therefore tends to shut the reactor down. Xenon-135 accumulation can be controlled by keeping power levels high enough to destroy it by neutron absorption as fast as it is produced. Fission also produces iodine-135, which in turn decays (with a half-life of 6.57 hours) to new xenon-135. When the reactor is shut down, iodine-135 continues to decay to xenon-135, making restarting the reactor more difficult for a day or two, as the xenon-135 decays into cesium-135, which is not nearly as poisonous as xenon-135, with a half-life of 9.2 hours. This temporary state is the "iodine pit." If the reactor has sufficient extra reactivity capacity, it can be restarted. As the extra xenon-135 is transmuted to xenon-136, which is much less a neutron poison, within a few hours the reactor experiences a "xenon burnoff (power) transient". Control rods must be further inserted to replace the neutron absorption of the lost xenon-135. Failure to properly follow such a procedure was a key step in the Chernobyl disaster.

Reactors used in nuclear marine propulsion (especially nuclear submarines) often cannot be run at continuous power around the clock in the same way that land-based power reactors are normally run, and in addition often need to have a very long core life without refueling. For this reason many designs use highly enriched uranium but incorporate burnable neutron poison in the fuel rods. This allows the reactor to be constructed with an excess of fissionable material, which is nevertheless made relatively safe early in the reactor's fuel burn-cycle by the presence of the neutron-absorbing material which is later replaced by normally produced long-lived neutron poisons (far longer-lived than xenon-135) which gradually accumulate over the fuel load's operating life.

Electrical Power Generation

The energy released in the fission process generates heat, some of which can be converted into usable energy. A common method of harnessing this thermal energy is to use it to boil water to produce pressurized steam which will then drive a steam turbine that turns an alternator and generates electricity.

Early Reactors

The Chicago Pile, the first nuclear reactor, built in secrecy at the University of Chicago in 1942 during World War II as part of the US's Manhattan project.

The neutron was discovered in 1932. The concept of a nuclear chain reaction brought about by nuclear reactions mediated by neutrons was first realized shortly thereafter, by Hungarian scientist Leó Szilárd, in 1933. He filed a patent for his idea of a simple reactor the following year while working at the Admiralty in London. However, Szilárd's idea did not incorporate the idea of nuclear fission as a neutron source, since that process was not yet discovered. Szilárd's ideas for nuclear reactors using neutron-mediated nuclear chain reactions in light elements proved unworkable.

Lise Meitner and Otto Hahn in their laboratory.

Inspiration for a new type of reactor using uranium came from the discovery by Lise Meitner, Fritz Strassmann and Otto Hahn in 1938 that bombardment of uranium with neutrons (provided by an alpha-on-beryllium fusion reaction, a "neutron howitzer") produced a barium residue, which they reasoned was created by the fissioning of the uranium nuclei. Subsequent studies in early 1939 (one of them by Szilárd and Fermi) revealed that several neutrons were also released during the fissioning, making available the opportunity for the nuclear chain reaction that Szilárd had envisioned six years previously.

On 2 August 1939 Albert Einstein signed a letter to President Franklin D. Roosevelt (written by Szilárd) suggesting that the discovery of uranium's fission could lead to the development of "extremely powerful bombs of a new type", giving impetus to the study of reactors and fission. Szilárd and Einstein knew each other well and had worked together years previously, but Einstein had never thought about this possibility for nuclear energy until Szilard reported it to him, at the beginning of his quest to produce the Einstein-Szilárd letter to alert the U.S. government.

Shortly after, Hitler's Germany invaded Poland in 1939, starting World War II in Europe. The U.S. was not yet officially at war, but in October, when the Einstein-Szilárd letter was delivered to him, Roosevelt commented that the purpose of doing the research was to make sure "the Nazis don't blow us up." The U.S. nuclear project followed, although with some delay as there remained skepticism (some of it from Fermi) and also little action from the small number of officials in the government who were initially charged with moving the project forward.

The following year the U.S. Government received the Frisch–Peierls memorandum from the UK, which stated that the amount of uranium needed for a chain reaction was far lower than had previously been thought. The memorandum was a product of the MAUD Committee, which was working on the UK atomic bomb project, known as Tube Alloys, later to be subsumed within the Manhattan Project.

The Chicago Pile Team, including Enrico Fermi and Leó Szilárd.

Eventually, the first artificial nuclear reactor, Chicago Pile-1, was constructed at the University of Chicago, by a team led by Enrico Fermi, in late 1942. By this time, the program had been pressured for a year by U.S. entry into the war. The Chicago Pile achieved criticality on 2 December 1942 at 3:25 PM. The reactor support structure was made of wood, which supported a pile (hence the name) of graphite blocks, embedded in which was natural uranium-oxide 'pseudospheres' or 'briquettes'.

Soon after the Chicago Pile, the U.S. military developed a number of nuclear reactors for the Manhattan Project starting in 1943. The primary purpose for the largest reactors (located at the

Hanford Site in Washington state), was the mass production of plutonium for nuclear weapons. Fermi and Szilard applied for a patent on reactors on 19 December 1944. Its issuance was delayed for 10 years because of wartime secrecy.

"World's first nuclear power plant" is the claim made by signs at the site of the EBR-I, which is now a museum near Arco, Idaho. Originally called "Chicago Pile-4", it was carried out under the direction of Walter Zinn for Argonne National Laboratory. This experimental LMFBR operated by the U.S. Atomic Energy Commission produced 0.8 kW in a test on 20 December 1951 and 100 kW (electrical) the following day, having a design output of 200 kW (electrical).

Besides the military uses of nuclear reactors, there were political reasons to pursue civilian use of atomic energy. U.S. President Dwight Eisenhower made his famous Atoms for Peace speech to the UN General Assembly on 8 December 1953. This diplomacy led to the dissemination of reactor technology to U.S. institutions and worldwide.

The first nuclear power plant built for civil purposes was the AM-1 Obninsk Nuclear Power Plant, launched on 27 June 1954 in the Soviet Union. It produced around 5 MW (electrical).

After World War II, the U.S. military sought other uses for nuclear reactor technology. Research by the Army and the Air Force never came to fruition; however, the U.S. Navy succeeded when they steamed the USS *Nautilus* (SSN-571) on nuclear power 17 January 1955.

The first commercial nuclear power station, Calder Hall in Sellafield, England was opened in 1956 with an initial capacity of 50 MW (later 200 MW).

The first portable nuclear reactor "Alco PM-2A" used to generate electrical power (2 MW) for Camp Century from 1960.

Components

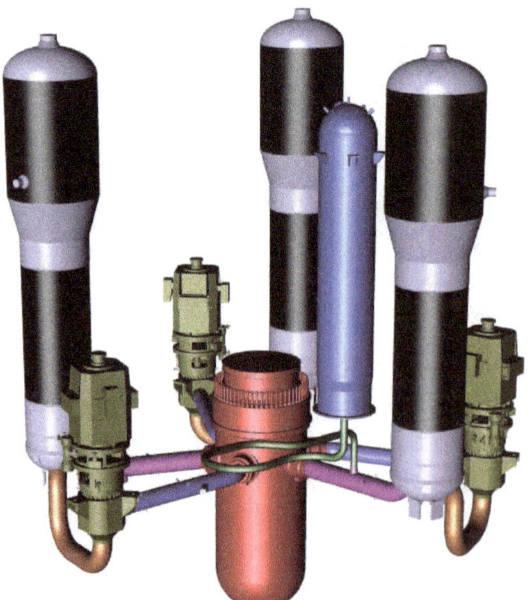

Primary coolant system showing reactor pressure vessel (red), steam generators (purple), pressurizer (blue), and pumps (green) in the three coolant loop Hualong One pressurized water reactor design

The key components common to most types of nuclear power plants are:

- Nuclear fuel
- Nuclear reactor core
- Neutron moderator
- Startup neutron source
- Neutron poison
- Neutron howitzer (provides steady source of neutrons to re-initiate reaction following shutdown)
- Coolant (often the Neutron Moderator and the Coolant are the same, usually both purified water)
- Control rods
- Reactor pressure vessel (RPV)
- Boiler feedwater pump
- Steam generators (not in BWRs)
- Steam turbine
- Electrical generator
- Condenser
- Cooling tower (not always required)
- Radwaste System (a section of the plant handling radioactive waste)
- Refueling Floor
- Spent fuel pool
- Nuclear safety systems
    - Reactor Protective System (RPS)
    - Emergency Diesel Generators
    - Emergency Core Cooling Systems (ECCS)
    - Standby Liquid Control System (emergency boron injection, in BWRs only)
- Essential service water system (ESWS)
- Containment building
- Control room

- Emergency Operations Facility
- Nuclear training facility (usually contains a Control Room simulator)

Reactor Types

NC State's PULSTAR Reactor is a 1 MW pool-type research reactor with 4% enriched, pin-type fuel consisting of UO_2 pellets in zircaloy cladding.

Classifications

Nuclear Reactors are classified by several methods; a brief outline of these classification methods is provided.

Classification by Type of Nuclear Reaction

Nuclear Fission

All commercial power reactors are based on nuclear fission. They generally use uranium and its product plutonium as nuclear fuel, though a thorium fuel cycle is also possible. Fission reactors can be divided roughly into two classes, depending on the energy of the neutrons that sustain the fission chain reaction:

- Thermal reactors (the most common type of nuclear reactor) use slowed or thermal neutrons to keep up the fission of their fuel. Almost all current reactors are of this type. These contain neutron moderator materials that slow neutrons until their neutron temperature is *thermalized*, that is, until their kinetic energy approaches the average kinetic energy of the surrounding particles. Thermal neutrons have a far higher cross-section (probability) of fissioning the fissile nuclei uranium-235, plutonium-239, and plutonium-241, and a relatively lower probability of neutron capture by uranium-238 (U-238) compared to the faster neutrons that originally result from fission, allowing use of low-enriched uranium or even natural uranium fuel. The moderator is often also the coolant, usually water under high pressure to increase the boiling point. These are surrounded by a reactor vessel, instrumentation to monitor and control the reactor, radiation shielding, and a containment building.

- Fast neutron reactors use fast neutrons to cause fission in their fuel. They do not have a neutron moderator, and use less-moderating coolants. Maintaining a chain reaction requires the fuel to be more highly enriched in fissile material (about 20% or more) due to the relatively lower probability of fission versus capture by U-238. Fast reactors have the potential to produce less transuranic waste because all actinides are fissionable with fast neutrons, but they are more difficult to build and more expensive to operate. Overall, fast reactors are less common than thermal reactors in most applications. Some early power stations were fast reactors, as are some Russian naval propulsion units. Construction of prototypes is continuing.

Nuclear Fusion

Fusion power is an experimental technology, generally with hydrogen as fuel. While not suitable for power production, Farnsworth-Hirsch fusors are used to produce neutron radiation.

Classification by Moderator Material

Used by thermal reactors:

- Graphite-moderated reactors
- Water moderated reactors
    - Heavy-water reactors (Used in Canada, India, Argentina, China, Pakistan, Romania and South Korea).
    - Light-water-moderated reactors (LWRs). Light-water reactors (the most common type of thermal reactor) use ordinary water to moderate and cool the reactors. When at operating temperature, if the temperature of the water increases, its density drops, and fewer neutrons passing through it are slowed enough to trigger further reactions. That negative feedback stabilizes the reaction rate. Graphite and heavy-water reactors tend to be more thoroughly thermalized than light water reactors. Due to the extra thermalization, these types can use natural uranium/unenriched fuel.
- Light-element-moderated reactors.
    - Molten salt reactors (MSRs) are moderated by light elements such as lithium or beryllium, which are constituents of the coolant/fuel matrix salts LiF and BeF_2.
    - Liquid metal cooled reactors, such as those whose coolant is a mixture of lead and bismuth, may use BeO as a moderator.
- Organically moderated reactors (OMR) use biphenyl and terphenyl as moderator and coolant.

Classification by Coolant

Treatment of the interior part of a VVER-1000 reactor frame on Atommash.

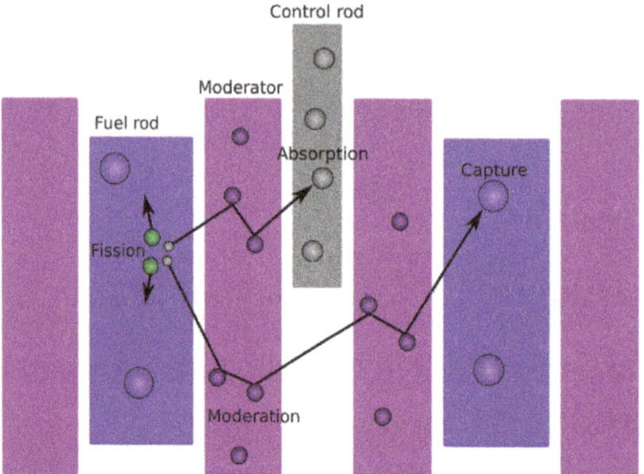

In thermal nuclear reactors (LWRs in specific), the coolant acts as a moderator that must slow down the neutrons before they can be efficiently absorbed by the fuel.

- Water cooled reactor. There are 104 operating reactors in the United States. Of these, 69 are pressurized water reactors (PWR), and 35 are boiling water reactors (BWR).
    - Pressurized water reactor (PWR) Pressurized water reactors constitute the large majority of all Western nuclear power plants.
        - A primary characteristic of PWRs is a pressurizer, a specialized pressure vessel. Most commercial PWRs and naval reactors use pressurizers. During normal operation, a pressurizer is partially filled with water, and a steam bubble is maintained above it by heating the water with submerged heaters. During normal operation, the pressurizer is connected to the primary reactor pressure vessel (RPV) and the pressurizer "bubble" provides an expansion space for changes in water volume in the reactor. This arrangement also provides a means of pressure control for the reactor by increasing or decreasing the steam pressure in the pressurizer using the pressurizer heaters.

- ☐ Pressurised heavy water reactors are a subset of pressurized water reactors, sharing the use of a pressurized, isolated heat transport loop, but using heavy water as coolant and moderator for the greater neutron economies it offers.

  - o Boiling water reactor (BWR)

    - ☐ BWRs are characterized by boiling water around the fuel rods in the lower portion of a primary reactor pressure vessel. A boiling water reactor uses ^{235}U, enriched as uranium dioxide, as its fuel. The fuel is assembled into rods housed in a steel vessel that is submerged in water. The nuclear fission causes the water to boil, generating steam. This steam flows through pipes into turbines. The turbines are driven by the steam, and this process generates electricity. During normal operation, pressure is controlled by the amount of steam flowing from the reactor pressure vessel to the turbine.

  - o Pool-type reactor

- Liquid metal cooled reactor. Since water is a moderator, it cannot be used as a coolant in a fast reactor. Liquid metal coolants have included sodium, NaK, lead, lead-bismuth eutectic, and in early reactors, mercury.

  - o Sodium-cooled fast reactor

  - o Lead-cooled fast reactor

- Gas cooled reactors are cooled by a circulating inert gas, often helium in high-temperature designs, while carbon dioxide has been used in past British and French nuclear power plants. Nitrogen has also been used. Utilization of the heat varies, depending on the reactor. Some reactors run hot enough that the gas can directly power a gas turbine. Older designs usually run the gas through a heat exchanger to make steam for a steam turbine.

- Molten salt reactors (MSRs) are cooled by circulating a molten salt, typically a eutectic mixture of fluoride salts, such as FLiBe. In a typical MSR, the coolant is also used as a matrix in which the fissile material is dissolved.

Classification by Generation

- Generation I reactor (early prototypes, research reactors, non-commercial power producing reactors)

- Generation II reactor (most current nuclear power plants 1965–1996)

- Generation III reactor (evolutionary improvements of existing designs 1996-now)

- Generation IV reactor (technologies still under development unknown start date, possibly 2030)

In 2003, the French Commissariat à l'Énergie Atomique (CEA) was the first to refer to "Gen II" types in Nucleonics Week.

The first mentioning of "Gen III" was in 2000, in conjunction with the launch of the Generation IV International Forum (GIF) plans.

"Gen IV" was named in 2000, by the United States Department of Energy (DOE) for developing new plant types.

Classification by Phase of Fuel

- Solid fueled
- Fluid fueled
  - Aqueous homogeneous reactor
  - Molten salt reactor
- Gas fueled (theoretical)

Classification by Use

- Electricity
  - Nuclear power plants including small modular reactors
- Propulsion
  - Nuclear marine propulsion
  - Various proposed forms of rocket propulsion
- Other uses of heat
  - Desalination
  - Heat for domestic and industrial heating
  - Hydrogen production for use in a hydrogen economy
- Production reactors for transmutation of elements
  - Breeder reactors are capable of producing more fissile material than they consume during the fission chain reaction (by converting fertile U-238 to Pu-239, or Th-232 to U-233). Thus, a uranium breeder reactor, once running, can be re-fueled with natural or even depleted uranium, and a thorium breeder reactor can be re-fueled with thorium; however, an initial stock of fissile material is required
  - Creating various radioactive isotopes, such as americium for use in smoke detectors, and cobalt-60, molybdenum-99 and others, used for imaging and medical treatment
  - Production of materials for nuclear weapons such as weapons-grade plutonium

- Providing a source of neutron radiation (for example with the pulsed Godiva device) and positron radiation (e.g. neutron activation analysis and potassium-argon dating)
- Research reactor: Typically reactors used for research and training, materials testing, or the production of radioisotopes for medicine and industry. These are much smaller than power reactors or those propelling ships, and many are on university campuses. There are about 280 such reactors operating, in 56 countries. Some operate with high-enriched uranium fuel, and international efforts are underway to substitute low-enriched fuel.

Current Technologies

Diablo Canyon — a PWR

Pressurized Water Reactors (PWR)

These reactors use a pressure vessel to contain the nuclear fuel, control rods, moderator, and coolant. They are cooled and moderated by high-pressure liquid water. The hot radioactive water that leaves the pressure vessel is looped through a steam generator, which in turn heats a secondary (non-radioactive) loop of water to steam that can run turbines. They are the majority of current reactors. This is a thermal neutron reactor design, the newest of which are the VVER-1200, Advanced Pressurized Water Reactor and the European Pressurized Reactor. United States Naval reactors are of this type.

Boiling Water Reactors (BWR)

A BWR is like a PWR without the steam generator. A boiling water reactor is cooled and moderated by water like a PWR, but at a lower pressure, which allows the water to boil inside the pressure vessel producing the steam that runs the turbines. Unlike a PWR, there is no primary and secondary loop. The thermal efficiency of these reactors can be higher, and they can be simpler, and even potentially more stable and safe. This is a thermal neutron reactor design, the newest of which are the Advanced Boiling Water Reactor and the Economic Simplified Boiling Water Reactor.

Pressurized Heavy Water Reactor (PHWR)

A Canadian design (known as CANDU), these reactors are heavy-water-cooled and -moderated pressurized-water reactors. Instead of using a single large pressure vessel as in a PWR, the fuel is contained in hundreds of pressure tubes. These reactors are fueled with natural uranium and are thermal neutron reactor designs. PHWRs can be refueled while at full power, which makes them very efficient in their use of uranium (it allows for precise flux control in the core). CANDU PHWRs have been built in Canada, Argentina, China, India, Pakistan, Romania, and South Korea. India also operates a number of PHWRs, often termed 'CANDU-derivatives', built after the Government of Canada halted nuclear dealings with India following the 1974 Smiling Buddha nuclear weapon test.

The CANDU Qinshan Nuclear Power Plant

Reaktor Bolshoy Moschnosti Kanalniy (High Power Channel Reactor) (RBMK)

A Soviet design, built to produce plutonium as well as power. RBMKs are water cooled with a graphite moderator. RBMKs are in some respects similar to CANDU in that they are refuelable during power operation and employ a pressure tube design instead of a PWR-style pressure vessel. However, unlike CANDU they are very unstable and large, making containment buildings for them expensive. A series of critical safety flaws have also been identified with the RBMK design, though some of these were corrected following the Chernobyl disaster. Their main attraction is their use of light water and un-enriched uranium. As of 2010, 11 remain open, mostly due to safety improvements and help from international safety agencies such as the DOE. Despite these safety improvements, RBMK reactors are still considered one of the most dangerous reactor designs in use. RBMK reactors were deployed only in the former Soviet Union.

The Ignalina Nuclear Power Plant — a RBMK type (closed 2009)

Gas-cooled Reactor (GCR) and Advanced Gas-cooled Reactor (AGR)

These are generally graphite moderated and CO_2 cooled. They can have a high thermal efficiency compared with PWRs due to higher operating temperatures. There are a number of operating reactors of this design, mostly in the United Kingdom, where the concept was developed. Older designs (i.e. Magnox stations) are either shut down or will be in the near future. However, the AGCRs have an anticipated life of a further 10 to 20 years. This is a thermal neutron reactor design. Decommissioning costs can be high due to large volume of reactor core.

The Magnox Sizewell A nuclear power station

The Torness nuclear power station — an AGR

Liquid-metal Fast-breeder Reactor (LMFBR)

This is a reactor design that is cooled by liquid metal, totally unmoderated, and produces more fuel than it consumes. They are said to "breed" fuel, because they produce fissionable fuel during operation because of neutron capture. These reactors can function much like a PWR in terms of efficiency, and do not require much high-pressure containment, as the liquid metal does not need to be kept at high pressure, even at very high temperatures. BN-350 and BN-600 in USSR and Superphénix in France were a reactor of this type, as was Fermi-I in the United States. The Monju reactor in Japan suffered a sodium leak in 1995 and was restarted in May 2010. All of them use/used liquid sodium. These reactors are fast neutron, not thermal neutron designs. These reactors come in two types:

The Superphénix, one of the few FBRs

Lead-cooled

Using lead as the liquid metal provides excellent radiation shielding, and allows for operation at very high temperatures. Also, lead is (mostly) transparent to neutrons, so fewer neutrons are lost in the coolant, and the coolant does not become radioactive. Unlike sodium, lead is mostly inert, so there is less risk of explosion or accident, but such large quantities of lead may be problematic from toxicology and disposal points of view. Often a reactor of this type would use a lead-bismuth eutectic mixture. In this case, the bismuth would present some minor radiation problems, as it is not quite as transparent to neutrons, and can be transmuted to a radioactive isotope more readily than lead. The Russian Alfa class submarine uses a lead-bismuth-cooled fast reactor as its main power plant.

Sodium-cooled

Most LMFBRs are of this type. The sodium is relatively easy to obtain and work with, and it also manages to actually prevent corrosion on the various reactor parts immersed in it. However, sodium explodes violently when exposed to water, so care must be taken, but such explosions would not be vastly more violent than (for example) a leak of superheated fluid from a SCWR or PWR. EBR-I, the first reactor to have a core meltdown, was of this type.

Pebble-bed Reactors (PBR)

These use fuel molded into ceramic balls, and then circulate gas through the balls. The result is an efficient, low-maintenance, very safe reactor with inexpensive, standardized fuel. The prototype was the AVR.

Molten Salt Reactors

These dissolve the fuels in fluoride salts, or use fluoride salts for coolant. These have many safety features, high efficiency and a high power density suitable for vehicles. Notably, they have no high pressures or flammable components in the core. The prototype was the MSRE, which also used the Thorium fuel cycle. As a breeder reactor type, it reprocesses the spent fuel, extracting both Uranium and transuranics, leaving only 0.1% of transuranic waste compared to conventional once-through uranium-fueled light water reactors currently in use. A separate issue are the radioactive fission products, which are not reprocessable and need to be disposed of as with conventional reactors.

Aqueous Homogeneous Reactor (AHR)

These reactors use soluble nuclear salts dissolved in water and mixed with a coolant and a neutron moderator.

Future and Developing Technologies

Advanced Reactors

More than a dozen advanced reactor designs are in various stages of development. Some are evolutionary from the PWR, BWR and PHWR designs above, some are more radical departures. The former include the advanced boiling water reactor (ABWR), two of which are now operating with others under construction, and the planned passively safe Economic Simplified Boiling Water Reactor (ESBWR) and AP1000 units.

- The Integral Fast Reactor (IFR) was built, tested and evaluated during the 1980s and then retired under the Clinton administration in the 1990s due to nuclear non-proliferation policies of the administration. Recycling spent fuel is the core of its design and it therefore produces only a fraction of the waste of current reactors.
- The pebble-bed reactor, a high-temperature gas-cooled reactor (HTGCR), is designed so high temperatures reduce power output by Doppler broadening of the fuel's neutron cross-section. It uses ceramic fuels so its safe operating temperatures exceed the power-reduction temperature range. Most designs are cooled by inert helium. Helium is not subject to steam explosions, resists neutron absorption leading to radioactivity, and does not dissolve contaminants that can become radioactive. Typical designs have more layers (up to 7) of passive containment than light water reactors (usually 3). A unique feature that may aid safety is that the fuel-balls actually form the core's mechanism, and are replaced one-by-one as they age. The design of the fuel makes fuel reprocessing expensive.
- The Small, sealed, transportable, autonomous reactor (SSTAR) is being primarily researched and developed in the US, intended as a fast breeder reactor that is passively safe and could be remotely shut down in case the suspicion arises that it is being tampered with.
- The Clean And Environmentally Safe Advanced Reactor (CAESAR) is a nuclear reactor concept that uses steam as a moderator – this design is still in development.
- The Reduced moderation water reactor builds upon the Advanced boiling water reactor(ABWR) that is presently in use, it is not a complete fast reactor instead using mostly epithermal neutrons, which are between thermal and fast neutrons in speed.
- The hydrogen-moderated self-regulating nuclear power module (HPM) is a reactor design emanating from the Los Alamos National Laboratory that uses uranium hydride as fuel.
- Subcritical reactors are designed to be safer and more stable, but pose a number of engineering and economic difficulties. One example is the Energy amplifier.
- Thorium-based reactors. It is possible to convert Thorium-232 into U-233 in reactors specially designed for the purpose. In this way, thorium, which is four times more abundant than uranium, can be used to breed U-233 nuclear fuel. U-233 is also believed to have favourable nuclear properties as compared to traditionally used U-235, including better neutron economy and lower production of long lived transuranic waste.
    - Advanced heavy-water reactor (AHWR)— A proposed heavy water moderated nuclear power reactor that will be the next generation design of the PHWR type. Under development in the Bhabha Atomic Research Centre (BARC), India.
    - KAMINI — A unique reactor using Uranium-233 isotope for fuel. Built in India by BARC and Indira Gandhi Center for Atomic Research (IGCAR).

- India is also planning to build fast breeder reactors using the thorium – Uranium-233 fuel cycle. The FBTR (Fast Breeder Test Reactor) in operation at Kalpakkam (India) uses Plutonium as a fuel and liquid sodium as a coolant.

Generation IV Reactors

Generation IV reactors are a set of theoretical nuclear reactor designs currently being researched. These designs are generally not expected to be available for commercial construction before 2030. Current reactors in operation around the world are generally considered second- or third-generation systems, with the first-generation systems having been retired some time ago. Research into these reactor types was officially started by the Generation IV International Forum (GIF) based on eight technology goals. The primary goals being to improve nuclear safety, improve proliferation resistance, minimize waste and natural resource utilization, and to decrease the cost to build and run such plants.

- Gas-cooled fast reactor
- Lead-cooled fast reactor
- Molten salt reactor
- Sodium-cooled fast reactor
- Supercritical water reactor
- Very-high-temperature reactor

Generation V+ Reactors

Generation V reactors are designs which are theoretically possible, but which are not being actively considered or researched at present. Though such reactors could be built with current or near term technology, they trigger little interest for reasons of economics, practicality, or safety.

- Liquid-core reactor. A closed loop liquid-core nuclear reactor, where the fissile material is molten uranium or uranium solution cooled by a working gas pumped in through holes in the base of the containment vessel.
- Gas-core reactor. A closed loop version of the nuclear lightbulb rocket, where the fissile material is gaseous uranium-hexafluoride contained in a fused silica vessel. A working gas (such as hydrogen) would flow around this vessel and absorb the UV light produced by the reaction. This reactor design could also function as a rocket engine, as featured in Harry Harrison's 1976 science-fiction novel 'Skyfall'. In theory, using UF_6 as a working fuel directly (rather than as a stage to one, as is done now) would mean lower processing costs, and very small reactors. In practice, running a reactor at such high power densities would probably produce unmanageable neutron flux, weakening most reactor materials, and therefore as the flux would be similar to that expected in fusion reactors, it would require similar materials to those selected by the International Fusion Materials Irradiation Facility.
    - Gas core EM reactor. As in the gas core reactor, but with photovoltaic arrays converting the UV light directly to electricity

- Fission fragment reactor
- Hybrid nuclear fusion. Would use the neutrons emitted by fusion to fission a blanket of fertile material, like U-238 or Th-232 and transmutate other reactor's spent nuclear fuel/ nuclear waste into relatively more benign isotopes.

Fusion Reactors

Controlled nuclear fusion could in principle be used in fusion power plants to produce power without the complexities of handling actinides, but significant scientific and technical obstacles remain. Several fusion reactors have been built, but only recently reactors have been able to release more energy than the amount of energy used in the process. Despite research having started in the 1950s, no commercial fusion reactor is expected before 2050. The ITER project is currently leading the effort to harness fusion power.

Nuclear Fuel Cycle

Thermal reactors generally depend on refined and enriched uranium. Some nuclear reactors can operate with a mixture of plutonium and uranium. The process by which uranium ore is mined, processed, enriched, used, possibly reprocessed and disposed of is known as the nuclear fuel cycle.

Under 1% of the uranium found in nature is the easily fissionable U-235 isotope and as a result most reactor designs require enriched fuel. Enrichment involves increasing the percentage of U-235 and is usually done by means of gaseous diffusion or gas centrifuge. The enriched result is then converted into uranium dioxide powder, which is pressed and fired into pellet form. These pellets are stacked into tubes which are then sealed and called fuel rods. Many of these fuel rods are used in each nuclear reactor.

Most BWR and PWR commercial reactors use uranium enriched to about 4% U-235, and some commercial reactors with a high neutron economy do not require the fuel to be enriched at all (that is, they can use natural uranium). According to the International Atomic Energy Agency there are at least 100 research reactors in the world fueled by highly enriched (weapons-grade/90% enrichment uranium). Theft risk of this fuel (potentially used in the production of a nuclear weapon) has led to campaigns advocating conversion of this type of reactor to low-enrichment uranium (which poses less threat of proliferation).

Fissile U-235 and non-fissile but fissionable and fertile U-238 are both used in the fission process. U-235 is fissionable by thermal (i.e. slow-moving) neutrons. A thermal neutron is one which is moving about the same speed as the atoms around it. Since all atoms vibrate proportionally to their absolute temperature, a thermal neutron has the best opportunity to fission U-235 when it is moving at this same vibrational speed. On the other hand, U-238 is more likely to capture a neutron when the neutron is moving very fast. This U-239 atom will soon decay into plutonium-239, which is another fuel. Pu-239 is a viable fuel and must be accounted for even when a highly enriched uranium fuel is used. Plutonium fissions will dominate the U-235 fissions in some reactors, especially after the initial loading of U-235 is spent. Plutonium is fissionable with both fast and thermal neutrons, which make it ideal for either nuclear reactors or nuclear bombs.

Most reactor designs in existence are thermal reactors and typically use water as a neutron moderator (moderator means that it slows down the neutron to a thermal speed) and as a coolant. But in a fast breeder reactor, some other kind of coolant is used which will not moderate or slow the neutrons down much. This enables fast neutrons to dominate, which can effectively be used to constantly replenish the fuel supply. By merely placing cheap unenriched uranium into such a core, the non-fissionable U-238 will be turned into Pu-239, "breeding" fuel.

In thorium fuel cycle thorium-232 absorbs a neutron in either a fast or thermal reactor. The thorium-233 beta decays to protactinium-233 and then to uranium-233, which in turn is used as fuel. Hence, like uranium-238, thorium-232 is a fertile material.

Fueling of Nuclear Reactors

The amount of energy in the reservoir of nuclear fuel is frequently expressed in terms of "full-power days," which is the number of 24-hour periods (days) a reactor is scheduled for operation at full power output for the generation of heat energy. The number of full-power days in a reactor's operating cycle (between refueling outage times) is related to the amount of fissile uranium-235 (U-235) contained in the fuel assemblies at the beginning of the cycle. A higher percentage of U-235 in the core at the beginning of a cycle will permit the reactor to be run for a greater number of full-power days.

At the end of the operating cycle, the fuel in some of the assemblies is "spent" and is discharged and replaced with new (fresh) fuel assemblies, although in practice it is the buildup of reaction poisons in nuclear fuel that determines the lifetime of nuclear fuel in a reactor. Long before all possible fission has taken place, the buildup of long-lived neutron absorbing fission byproducts impedes the chain reaction. The fraction of the reactor's fuel core replaced during refueling is typically one-fourth for a boiling-water reactor and one-third for a pressurized-water reactor. The disposition and storage of this spent fuel is one of the most challenging aspects of the operation of a commercial nuclear power plant. This nuclear waste is highly radioactive and its toxicity presents a danger for thousands of years.

Not all reactors need to be shut down for refueling; for example, pebble bed reactors, RBMK reactors, molten salt reactors, Magnox, AGR and CANDU reactors allow fuel to be shifted through the reactor while it is running. In a CANDU reactor, this also allows individual fuel elements to be situated within the reactor core that are best suited to the amount of U-235 in the fuel element.

The amount of energy extracted from nuclear fuel is called its burnup, which is expressed in terms of the heat energy produced per initial unit of fuel weight. Burn up is commonly expressed as megawatt days thermal per metric ton of initial heavy metal.

Nuclear Safety Concerns and Controversy

Nuclear safety covers the actions taken to prevent nuclear and radiation accidents or to limit their consequences. The nuclear power industry has improved the safety and performance of reactors, and has proposed new safer (but generally untested) reactor designs but there is no guarantee that the reactors will be designed, built and operated correctly. Mistakes do occur and the designers of reactors at Fukushima in Japan did not anticipate that a tsunami generated by an earthquake would disable the backup systems that were supposed to stabilize the reactor after the earthquake, despite multiple warnings by the NRG and the Japanese nuclear safety administration. According to UBS AG, the Fukushima I nuclear accidents have cast doubt on whether even an advanced economy like Japan can master nuclear safety. Catastrophic scenarios involving terrorist attacks are also conceivable. An interdisciplinary team from MIT has estimated that given the expected growth of nuclear power from 2005–2055, at least four serious nuclear accidents would be expected in that period.

Nuclear Accidents and Controversy

Three of the reactors at Fukushima I overheated, causing meltdowns that eventually led to explosions, which released large amounts of radioactive material into the air.

Some serious nuclear and radiation accidents have occurred. Nuclear power plant accidents include the SL-1 accident (1961), the Three Mile Island accident (1979), Chernobyl disaster (1986), and the Fukushima Daiichi nuclear disaster (2011). Nuclear-powered submarine mishaps include the K-19 reactor accident (1961), the K-27 reactor accident (1968), and the K-431 reactor accident (1985).

Nuclear reactors have been launched into Earth orbit at least 34 times. A number of incidents connected with the unmanned nuclear-reactor-powered Soviet RORSAT radar satellite program resulted in spent nuclear fuel re-entering the Earth's atmosphere from orbit.

Natural Nuclear Reactors

Although nuclear fission reactors are often thought of as being solely a product of modern technology, the first nuclear fission reactors were in fact naturally occurring. A natural nuclear fission reactor can occur under certain circumstances that mimic the conditions in a constructed reactor. Fifteen natural fission reactors have so far been found in three separate ore deposits at the Oklo uranium mine in Gabon, West Africa. First discovered in 1972 by French physicist Francis Perrin, they are collectively known as the Oklo Fossil Reactors. Self-sustaining nuclear fission reactions took place in these reactors approximately 1.5 billion years ago, and ran for a few hundred thousand years, averaging 100 kW of power output during that time. The concept of a natural nuclear reactor was theorized as early as 1956 by Paul Kuroda at the University of Arkansas.

Such reactors can no longer form on Earth: radioactive decay over this immense time span has reduced the proportion of U-235 in naturally occurring uranium to below the amount required to sustain a chain reaction.

The natural nuclear reactors formed when a uranium-rich mineral deposit became inundated with groundwater that acted as a neutron moderator, and a strong chain reaction took place. The water moderator would boil away as the reaction increased, slowing it back down again and preventing a meltdown. The fission reaction was sustained for hundreds of thousands of years.

These natural reactors are extensively studied by scientists interested in geologic radioactive waste disposal. They offer a case study of how radioactive isotopes migrate through the Earth's crust. This is a significant area of controversy as opponents of geologic waste disposal fear that isotopes from stored waste could end up in water supplies or be carried into the environment.

Emission

Nuclear reactor emits tiny amount of tritium, Sr-90 to air and groundwater. Water contaminated with tritium is colorless and odorless. Large doses of Sr-90 increases the risk of bone cancer and leukemia in animals, and is presumed to do so in people.

References

- "Oklo: Natural Nuclear Reactors". Office of Civilian Radioactive Waste Management. Archived from the original on 16 March 2006. Retrieved 28 June 2006
- "International Scientific Journal for Alternative Energy and Ecology, DIRECT CONVERSION OF NUCLEAR ENERGY TO ELECTRICITY, Mark A. Prelas" (PDF)
- Juhasz, Albert J.; Rarick, Richard A.; Rangarajan, Rajmohan. "High Efficiency Nuclear Power Plants Using Liquid Fluoride Thorium Reactor Technology" (PDF). NASA. Retrieved 27 October 2014
- Golubev, V. I.; Dolgov, V. V.; Dulin, V. A.; Zvonarev, A. V.; Smetanin, É. Y.; Kochetkov, L. A.; Korobeinikov, V. V.; Liforov, V. G.; Manturov, G. N.; Matveenko, I. P.; Tsibulya, A. M. (1993). "Fast-reactor actinoid transmutation". Atomic Energy. 74: 83. doi:10.1007/BF00750983
- "U.S. Nuclear Power Plants. General Statistical Information". Nuclear Energy Institute. Archived from the original on 22 October 2008. Retrieved 3 October 2009

2

Classification and Selection of Nuclear Fuel

The fuel used in nuclear power stations to produce heat required for turbines is known as nuclear fuel. Nuclear fuel cycle is the process of refining and purifying nuclear fuel. The chapter strategically encompasses and incorporates the major components and key concepts of nuclear science and technology, providing a complete understanding.

Nuclear Fuel

Nuclear fuel is a substance that is used in nuclear power stations to produce heat to power turbines. Heat is created when nuclear fuel undergoes nuclear fission.

Most nuclear fuels contain heavy fissile elements that are capable of nuclear fission, such as uranium-235 or plutonium-239. When the unstable nuclei of these atoms are hit by a slow-moving neutron, they split, creating two daughter nuclei and two or three more neutrons. These neutrons then go on to split more nuclei. This creates a self-sustaining chain reaction that is controlled in a nuclear reactor, or uncontrolled in a nuclear weapon.

The processes involved in mining, refining, purifying, using, and disposing of nuclear fuel are collectively known as the nuclear fuel cycle.

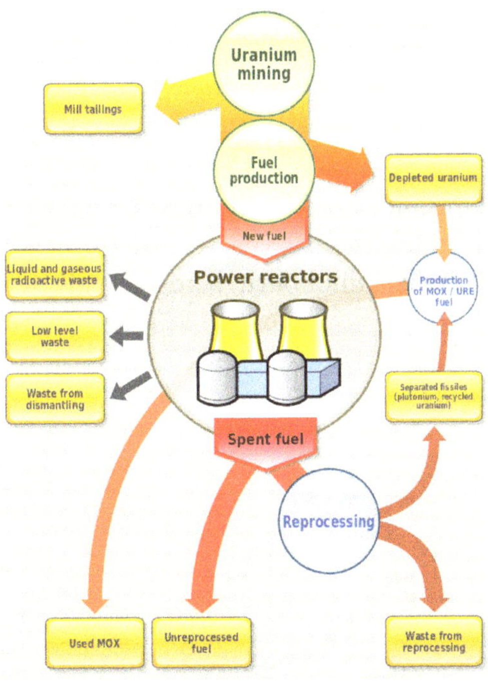

Nuclear Fuel Process

Not all types of nuclear fuels create power from nuclear fission; plutonium-238 and some other elements are used to produce small amounts of nuclear power by radioactive decay in radioisotope thermoelectric generators and other types of atomic batteries.

Nuclear fuel has the highest energy density of all practical fuel sources.

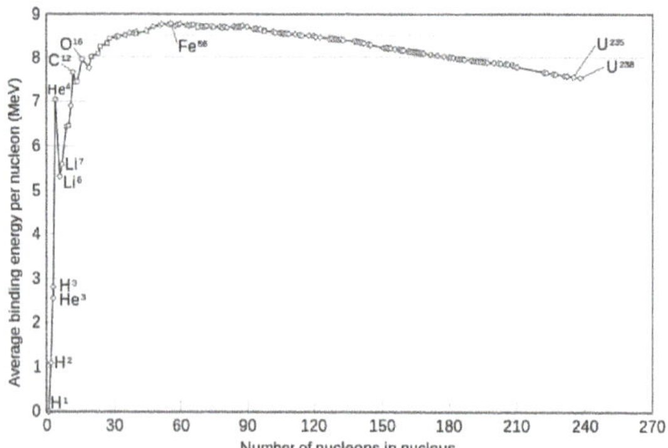

A graph comparing nucleon number against binding energy

Close-up of a replica of the core of the research reactor at the Institut Laue-Langevin

Oxide Fuel

For fission reactors, the fuel (typically based on uranium) is usually based on the metal oxide; the oxides are used rather than the metals themselves because the oxide melting point is much higher than that of the metal and because it cannot burn, being already in the oxidized state.

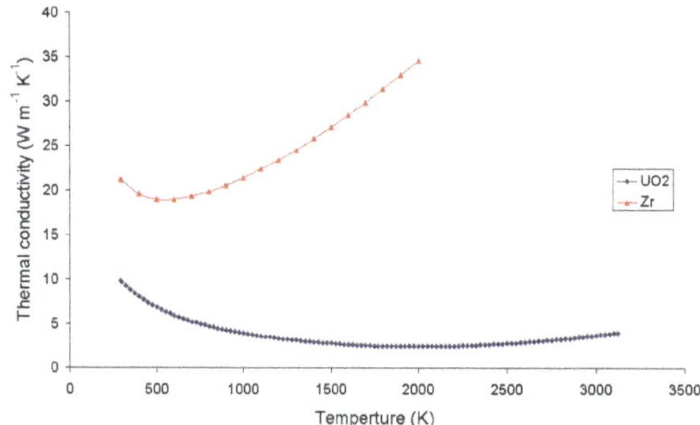

The thermal conductivity of zirconium metal and uranium dioxide as a function of temperature

UOX

Uranium dioxide is a black semiconducting solid. It can be made by reacting uranyl nitrate with a base (ammonia) to form a solid (ammonium uranate). It is heated (calcined) to form U_3O_8 that can then be converted by heating in an argon / hydrogen mixture (700 °C) to form UO_2. The UO_2 is then mixed with an organic binder and pressed into pellets, these pellets are then fired at a much higher temperature (in H_2/Ar) to sinter the solid. The aim is to form a dense solid which has few pores.

The thermal conductivity of uranium dioxide is very low compared with that of zirconium metal, and it goes down as the temperature goes up.

It is important to note that the corrosion of uranium dioxide in an aqueous environment is controlled by similar electrochemical processes to the galvanic corrosion of a metal surface.

MOX

Mixed oxide, or MOX fuel, is a blend of plutonium and natural or depleted uranium which behaves similarly (though not identically) to the enriched uranium feed for which most nuclear reactors were designed. MOX fuel is an alternative to low enriched uranium (LEU) fuel used in the light water reactors which predominate nuclear power generation.

Some concern has been expressed that used MOX cores will introduce new disposal challenges, though MOX is itself a means to dispose of surplus plutonium by transmutation.

Reprocessing of commercial nuclear fuel to make MOX was done in the Sellafield MOX Plant (England). As of 2015, MOX fuel is made in France), and to a lesser extent in Russia, India and Japan. China plans to develop fast breeder reactors and reprocessing.

The Global Nuclear Energy Partnership, was a U.S. proposal in the George W. Bush Administration to form an international partnership to see spent nuclear fuel reprocessed in a way that renders the plutonium in it usable for nuclear fuel but not for nuclear weapons. Reprocessing of spent commercial-reactor nuclear fuel has not been permitted in the United States due to nonproliferation considerations. All of the other reprocessing nations have long had nuclear weapons from military-focused "research"-reactor fuels except for Japan.Normally, with the fuel being changed every three years or so, about half of the Pu-239 is 'burned' in the reactor, providing about one third of the total energy. It behaves like U-235 and its fission releases a similar amount of energy. The higher the burn-up, the more plutonium in the spent fuel, but the lower the fraction of fissile plutonium. Typically about one percent of the used fuel discharged from a reactor is plutonium, and some two thirds of this is fissile (c. 50% Pu-239, 15% Pu-241). Worldwide, some 70 tonnes of plutonium contained in used fuel is removed when refueling reactors each year.

Metal Fuel

Metal fuels have the advantage of a much higher heat conductivity than oxide fuels but cannot survive equally high temperatures. Metal fuels have a long history of use, stretching from the Clementine reactor in 1946 to many test and research reactors. Metal fuels have the potential for the highest fissile atom density. Metal fuels are normally alloyed, but some metal fuels have been made with pure uranium metal. Uranium alloys that have been used include uranium aluminum, uranium zirconium, uranium silicon, uranium molybdenum, and uranium zirconium hydride. Any of the aforementioned fuels can be made with plutonium and other actinides as part of a closed nuclear fuel cycle. Metal fuels have been used in water reactors and liquid metal fast breeder reactors, such as EBR-II.

TRIGA Fuel

TRIGA fuel is used in TRIGA (Training, Research, Isotopes, General Atomics) reactors. The TRIGA reactor uses UZrH fuel, which has a prompt negative fuel temperature coefficient of reactivity, meaning that as the temperature of the core increases, the reactivity decreases—so it is highly unlikely for a meltdown to occur. Most cores that use this fuel are "high leakage" cores where the excess leaked neutrons can be utilized for research. TRIGA fuel was originally designed to use highly enriched uranium, however in 1978 the U.S. Department of Energy launched its Reduced Enrichment for Research Test Reactors program, which promoted reactor conversion to low-enriched uranium fuel. A total of 35 TRIGA reactors have been installed at locations across the USA. A further 35 reactors have been installed in other countries.

Actinide Fuel

In a fast neutron reactor, the minor actinides produced by neutron capture of uranium and plutonium can be used as fuel. Metal actinide fuel is typically an alloy of zirconium, uranium, plutonium, and minor actinides. It can be made inherently safe as thermal expansion of the metal alloy will increase neutron leakage.

Molten Plutonium

Molten plutonium, alloyed with other metals to lower its melting point and encapsulated in tantalum, was tested in two experimental reactors, LAMPRE I and LAMPRE II, at LANL in the 1960s. "LAMPRE experienced three separate fuel failures during operation."

Ceramic Fuels

Ceramic fuels other than oxides have the advantage of high heat conductivities and melting points, but they are more prone to swelling than oxide fuels and are not understood as well.

Uranium Nitride

This is often the fuel of choice for reactor designs that NASA produces, one advantage is that UN has a better thermal conductivity than UO_2. Uranium nitride has a very high melting point. This fuel has the disadvantage that unless ^{15}N was used (in place of the more common ^{14}N) that a large amount of ^{14}C would be generated from the nitrogen by the (n,p) reaction. As the nitrogen required for such a fuel would be so expensive it is likely that the fuel would have to be reprocessed by pyroprocessing to enable the ^{15}N to be recovered. It is likely that if the fuel was processed and dissolved in nitric acid that the nitrogen enriched with ^{15}N would be diluted with the common ^{14}N.

Uranium Carbide

Much of what is known about uranium carbide is in the form of pin-type fuel elements for liquid metal fast reactors during their intense study during the 1960s and 1970s. However, recently there has been a revived interest in uranium carbide in the form of plate fuel and most notably, micro fuel particles (such as TRISO particles).

The high thermal conductivity and high melting point makes uranium carbide an attractive fuel. In addition, because of the absence of oxygen in this fuel (during the course of irradiation, excess gas pressure can build from the formation of O_2 or other gases) as well as the ability to complement a ceramic coating (a ceramic-ceramic interface has structural and chemical advantages), uranium carbide could be the ideal fuel candidate for certain Generation IV reactors such as the gas-cooled fast reactor.

Liquid Fuels

Liquid fuels are liquids containing dissolved nuclear fuel and have been shown to offer numerous operational advantages compared to traditional solid fuel approaches.

Liquid-fuel reactors offer significant safety advantages due to their inherently stable "self-adjusting" reactor dynamics. This provides two major benefits: - virtually eliminating the possibility of a run-away reactor meltdown, - providing an automatic load-following capability which is well suited to electricity generation and high temperature industrial heat applications.

Another major advantage of the liquid core is its ability to be drained rapidly into a passively safe dump-tank. This advantage was conclusively demonstrated repeatedly as part of a weekly shutdown procedure during the highly successful 4 year ORNL MSRE program.

Another huge advantage of the liquid core is its ability to release xenon gas which normally acts as a neutron absorber and causes structural occlusions in solid fuel elements (leading to early replacement of solid fuel rods with over 98% of the nuclear fuel unburned, including many long lived actinides). In contrast Molten Salt Reactors (MSR) are capable of retaining the fuel mixture for significantly extended periods, which not only increases fuel efficiency dramatically, but also incinerates the vast majority of its own waste as part of the normal operational characteristics.

Molten Salts

Molten salt fuels have nuclear fuel dissolved directly in the molten salt coolant. Molten salt-fueled reactors, such as the liquid fluoride thorium reactor (LFTR), are different from molten salt-cooled reactors that do not dissolve nuclear fuel in the coolant.

Molten salt fuels were used in the LFTR known as the Molten Salt Reactor Experiment, as well as other liquid core reactor experiments. The liquid fuel for the molten salt reactor was a mixture of lithium, beryllium, thorium and uranium fluorides: $LiF\text{-}BeF_2\text{-}ThF_4\text{-}UF_4$ (72-16-12-0.4 mol%). It had a peak operating temperature of 705 °C in the experiment, but could have operated at much higher temperatures, since the boiling point of the molten salt was in excess of 1400 °C.

Aqueous Solutions of Uranyl Salts

The aqueous homogeneous reactors (AHRs) use a solution of uranyl sulfate or other uranium salt in water. Historically, AHRs have all been small research reactors, not large power reactors. An AHR known as the Medical Isotope Production System is being considered for production of medical isotopes.

Common Physical Forms of Nuclear Fuel

Uranium dioxide (UO_2) powder is compacted to cylindrical pellets and sintered at high temperatures to produce ceramic nuclear fuel pellets with a high density and well defined physical properties and chemical composition. A grinding process is used to achieve a uniform cylindrical geometry with narrow tolerances. Such fuel pellets are then stacked and filled into the metallic tubes. The metal used for the tubes depends on the design of the reactor. Stainless steel was used in the past, but most reactors now use a zirconium alloy which, in addition to being highly corrosion-resistant, has low neutron absorption. The tubes containing the fuel pellets are sealed: these tubes are called fuel rods. The finished fuel rods are grouped into fuel assemblies that are used to build up the core of a power reactor.

NRC photo of fresh fuel pellets ready for assembly.

Cladding is the outer layer of the fuel rods, standing between the coolant and the nuclear fuel. It is made of a corrosion-resistant material with low absorption cross section for thermal neutrons, usually Zircaloy or steel in modern constructions, or magnesium with small amount of aluminium and other metals for the now-obsolete Magnox reactors. Cladding prevents radioactive fission fragments from escaping the fuel into the coolant and contaminating it.

NRC photo of fresh fuel being inspected.

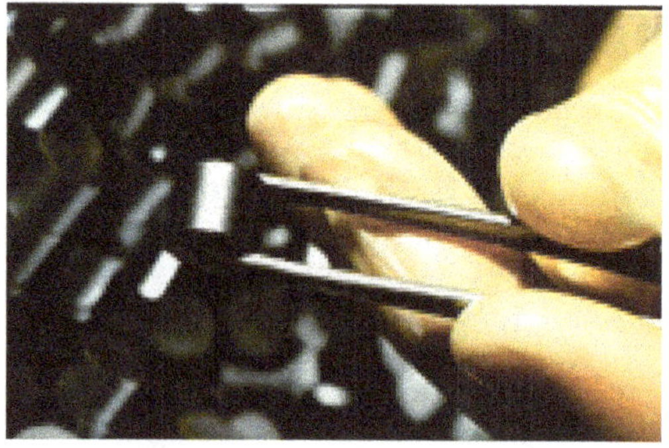

Nuclear Regulatory Commission (NRC) photo of unirradiated (fresh) fuel pellets.

PWR Fuel

Pressurized water reactor (PWR) fuel consists of cylindrical rods put into bundles. A uranium oxide ceramic is formed into pellets and inserted into Zircaloy tubes that are bundled together. The Zircaloy tubes are about 1 cm in diameter, and the fuel cladding gap is filled with helium gas to improve the conduction of heat from the fuel to the cladding. There are about 179-264 fuel rods per fuel bundle and about 121 to 193 fuel bundles are loaded into a reactor core. Generally, the fuel bundles consist of fuel rods bundled 14×14 to 17×17. PWR fuel bundles are about 4 meters long. In PWR fuel bundles, control rods are inserted through the top directly into the fuel bundle. The fuel bundles usually are enriched several percent in ^{235}U. The uranium oxide is dried before inserting into the tubes to try to eliminate moisture in the ceramic fuel that can lead to corrosion and hydrogen embrittlement. The Zircaloy tubes are pressurized with helium to try to minimize pellet-cladding interaction which can lead to fuel rod failure over long periods.

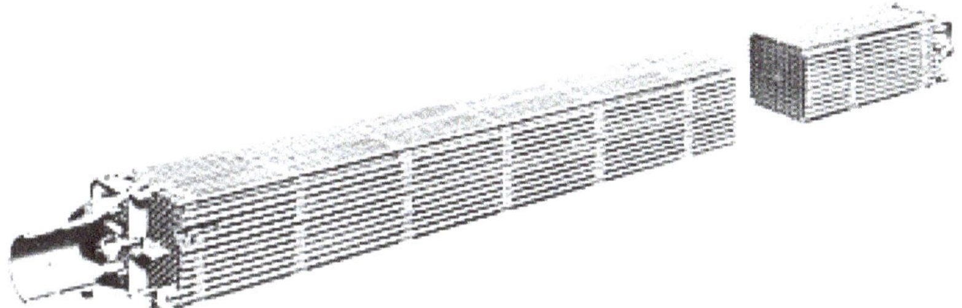

PWR fuel assembly (also known as a fuel bundle) This fuel assembly is from a pressurized water reactor of the nuclear-powered passenger and cargo ship NS Savannah. Designed and built by the Babcock & Wilcox Company.

BWR Fuel

In boiling water reactors (BWR), the fuel is similar to PWR fuel except that the bundles are "canned". That is, there is a thin tube surrounding each bundle. This is primarily done to prevent local density variations from affecting neutronics and thermal hydraulics of the reactor core. In modern BWR fuel bundles, there are either 91, 92, or 96 fuel rods per assembly depending on the manufacturer. A range between 368 assemblies for the smallest and 800 assemblies for the largest U.S. BWR forms the reactor core. Each BWR fuel rod is backfilled with helium to a pressure of about three atmospheres (300 kPa).

CANDU Fuel

CANDU fuel bundles are about a half meter long and 10 cm in diameter. They consist of sintered (UO_2) pellets in zirconium alloy tubes, welded to zirconium alloy end plates. Each bundle is roughly 20 kg, and a typical core loading is on the order of 4500-6500 bundles, depending on the design. Modern types typically have 37 identical fuel pins radially arranged about the long axis of the bundle, but in the past several different configurations and numbers of pins have been used. The CANFLEX bundle has 43 fuel elements, with two element sizes. It is also about 10 cm (4 inches) in diameter, 0.5 m (20 in) long and weighs about 20 kg (44 lb) and replaces the 37-pin standard bundle. It has been designed specifically to increase fuel performance by utilizing two different pin diameters. Current CANDU designs do not need enriched uranium to achieve criticality (due to their more efficient heavy water moderator), however, some newer concepts call for low enrichment to help reduce the size of the reactors.

CANDU fuel bundles Two CANDU ("CANada Deuterium Uranium") fuel bundles, each about 50 cm long, 10 cm in diameter.

Less-common Fuel Forms

Various other nuclear fuel forms find use in specific applications, but lack the widespread use of those found in BWRs, PWRs, and CANDU power plants. Many of these fuel forms are only found in research reactors, or have military applications.

Magnox Fuel

Magnox reactors are pressurised, carbon dioxide–cooled, graphite-moderated reactors using natural uranium (i.e. unenriched) as fuel and Magnox alloy as fuel cladding. Working pressure varies from 6.9 to 19.35 bar for the steel pressure vessels, and the two reinforced concrete designs operated at 24.8 and 27 bar. Magnox alloy consists mainly of magnesium with small amounts of aluminium and other metals—used in cladding unenriched uranium metal fuel with a non-oxidising covering to contain fission products. Magnox is short for Magnesium non-oxidising. This material has the advantage of a low neutron capture cross-section, but has two major disadvantages:

- It limits the maximum temperature, and hence the thermal efficiency, of the plant.
- It reacts with water, preventing long-term storage of spent fuel under water.

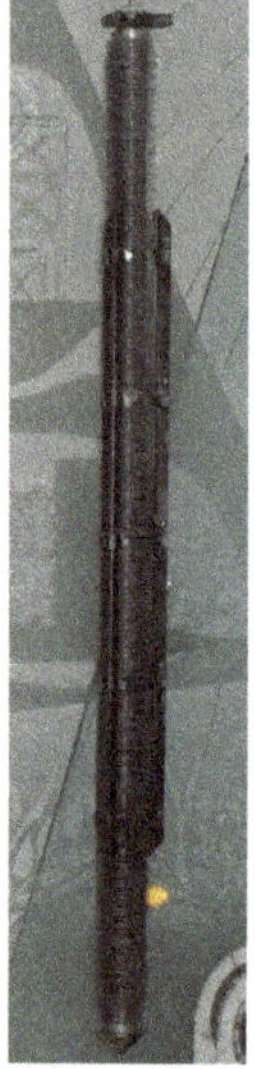

A magnox fuel rod

Magnox fuel incorporated cooling fins to provide maximum heat transfer despite low operating temperatures, making it expensive to produce. While the use of uranium metal rather than oxide made reprocessing more straightforward and therefore cheaper, the need to reprocess fuel a short time after removal from the reactor meant that the fission product hazard was severe. Expensive remote handling facilities were required to address this danger.

TRISO Fuel

Tristructural-isotropic (TRISO) fuel is a type of micro fuel particle. It consists of a fuel kernel composed of UO_x (sometimes UC or UCO) in the center, coated with four layers of three isotropic materials. The four layers are a porous buffer layer made of carbon, followed by a dense inner layer of pyrolytic carbon (PyC), followed by a ceramic layer of SiC to retain fission products at elevated temperatures and to give the TRISO particle more structural integrity, followed by a dense outer layer of PyC. TRISO fuel particles are designed not to crack due to the stresses from processes (such as differential thermal expansion or fission gas pressure) at temperatures up to and beyond 1600 °C, and therefore can contain the fuel in the worst of accident scenarios in a properly designed reactor. Two such reactor designs are the pebble-bed reactor (PBR), in which thousands of TRISO fuel particles are dispersed into graphite pebbles, and the prismatic-block gas-cooled reactor (such as the GT-MHR), in which the TRISO fuel particles are fabricated into compacts and placed in a graphite block matrix. Both of these reactor designs are high temperature gas reactors (HTGRs). These are also the basic reactor designs of very-high-temperature reactors (VHTRs), one of the six classes of reactor designs in the Generation IV initiative that is attempting to reach even higher HTGR outlet temperatures.

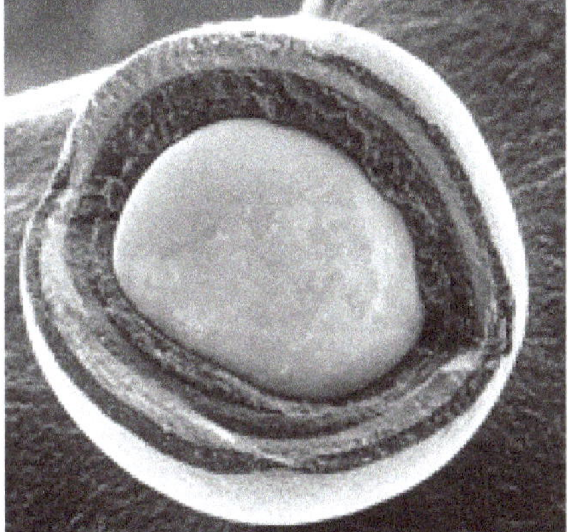

TRISO fuel particle which has been cracked, showing the multiple coating layers

TRISO fuel particles were originally developed in the United Kingdom as part of the Dragon reactor project. The inclusion of the SiC as diffusion barrier was first suggested by D. T. Livey. The first nuclear reactor to use TRISO fuels was the Dragon reactor and the first powerplant was the THTR-300. Currently, TRISO fuel compacts are being used in the experimental reactors, the HTR-10 in China, and the HTTR in Japan.

QUADRISO Fuel

In QUADRISO particles a burnable neutron poison (europium oxide or erbium oxide or carbide) layer surrounds the fuel kernel of ordinary TRISO particles to better manage the excess of reactivity. If the core is equipped both with TRISO and QUADRISO fuels, at beginning of life neutrons do not reach the fuel of the QUADRISO particles because they are stopped by the burnable poison. After irradiation, the poison depletes and neutrons stream into the fuel kernel of QUADRISO particles inducing fission reactions. This mechanism compensates fuel depletion of ordinary TRISO fuel. In the generalized QUADRISO fuel concept the poison can eventually be mixed with the fuel kernel or the outer pyrocarbon. The QUADRISO concept has been conceived at Argonne National Laboratory.

QUADRISO Particle

RBMK Fuel

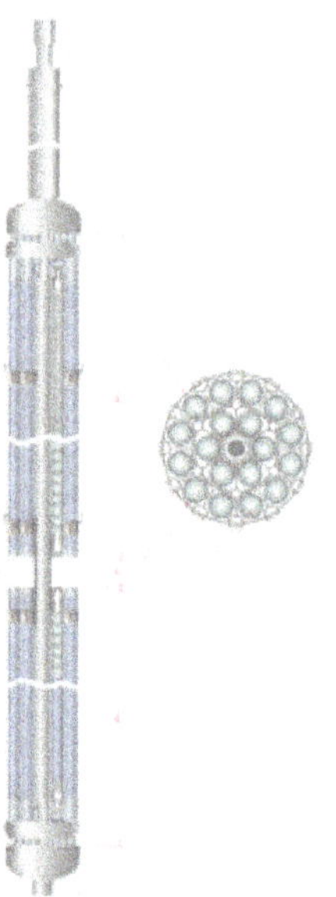

RBMK reactor fuel rod holder 1 – distancing armature; 2 – fuel rods shell; 3 – fuel tablets.

RBMK reactor fuel was used in Soviet-designed and built RBMK-type reactors. This is a low-enriched uranium oxide fuel. The fuel elements in an RBMK are 3 m long each, and two of these sit back-to-back on each fuel channel, pressure tube. Reprocessed uranium from Russian VVER reactor spent fuel is used to fabricate RBMK fuel. Following the Chernobyl accident, the enrichment of fuel was changed from 2.0% to 2.4%, to compensate for control rod modifications and the introduction of additional absorbers.

CerMet Fuel

CerMet fuel consists of ceramic fuel particles (usually uranium oxide) embedded in a metal matrix. It is hypothesized that this type of fuel is what is used in United States Navy reactors. This fuel has high heat transport characteristics and can withstand a large amount of expansion.

ATR Core The Advanced Test Reactor at Idaho National Laboratory uses plate-type fuel in a clover leaf arrangement. The blue glow around the core is known as Cherenkov radiation.

Plate-type Fuel

Plate-type fuel has fallen out of favor over the years. Plate-type fuel is commonly composed of enriched uranium sandwiched between metal cladding. Plate-type fuel is used in several research reactors where a high neutron flux is desired, for uses such as material irradiation studies or isotope production, without the high temperatures seen in ceramic, cylindrical fuel. It is currently used in the Advanced Test Reactor (ATR) at Idaho National Laboratory, and the nuclear research reactor at the University of Massachusetts Lowell Radiation Laboratory.

Sodium-bonded Fuel

Sodium-bonded fuel consists of fuel that has liquid sodium in the gap between the fuel slug (or pellet) and the cladding. This fuel type is often used for sodium-cooled liquid metal fast reactors. It has been used in EBR-I, EBR-II, and the FFTF. The fuel slug may be metallic or ceramic. The sodium bonding is used to reduce the temperature of the fuel.

Spent Nuclear Fuel

Used nuclear fuel is a complex mixture of the fission products, uranium, plutonium, and the transplutonium metals. In fuel which has been used at high temperature in power reactors it is common for the fuel to be *heterogeneous*; often the fuel will contain nanoparticles of platinum group metals such as palladium. Also the fuel may well have cracked, swollen, and been heated close to its melting point. Despite the fact that the used fuel can be cracked, it is very insoluble in water, and is able to retain the vast majority of the actinides and fission products within the uranium dioxide crystal lattice.

Oxide Fuel Under Accident Conditions

Two main modes of release exist, the fission products can be vaporised or small particles of the fuel can be dispersed.

Fuel Behavior and Post-irradiation Examination

Post-Irradiation Examination (PIE) is the study of used nuclear materials such as nuclear fuel. It has several purposes. It is known that by examination of used fuel that the failure modes which occur during normal use (and the manner in which the fuel will behave during an accident) can be studied. In addition information is gained which enables the users of fuel to assure themselves of its quality and it also assists in the development of new fuels. After major accidents the core (or what is left of it) is normally subject to PIE to find out what happened. One site where PIE is done is the ITU which is the EU centre for the study of highly radioactive materials.

Materials in a high-radiation environment (such as a reactor) can undergo unique behaviors such as swelling and non-thermal creep. If there are nuclear reactions within the material (such as what happens in the fuel), the stoichiometry will also change slowly over time. These behaviors can lead to new material properties, cracking, and fission gas release.

The thermal conductivity of uranium dioxide is low; it is affected by porosity and burn-up. The burn-up results in fission products being dissolved in the lattice (such as lanthanides), the precipitation of fission products such as palladium, the formation of fission gas bubbles due to fission products such as xenon and krypton and radiation damage of the lattice. The low thermal conductivity can lead to overheating of the center part of the pellets during use. The porosity results in a decrease in both the thermal conductivity of the fuel and the swelling which occurs during use.

According to the International Nuclear Safety Center the thermal conductivity of uranium dioxide can be predicted under different conditions by a series of equations.

The bulk density of the fuel can be related to the thermal conductivity

Where ρ is the bulk density of the fuel and ρ_{td} is the theoretical density of the uranium dioxide.

Then the thermal conductivity of the porous phase (K_f) is related to the conductivity of the perfect phase (K_o, no porosity) by the following equation. Note that s is a term for the shape factor of the holes.

$$K_f = K_o\left(1 - p / 1 + (s - 1)p\right)$$

Rather than measuring the thermal conductivity using the traditional methods such as Lees' disk, the Forbes' method, or Searle's bar, it is common to use Laser Flash Analysis where a small disc of fuel is placed in a furnace. After being heated to the required temperature one side of the disc is illuminated with a laser pulse, the time required for the heat wave to flow through the disc, the density of the disc, and the thickness of the disk can then be used to calculate and determine the thermal conductivity.

$$\lambda = \rho C_p \alpha$$

- λ thermal conductivity
- ρ density
- C_p heat capacity
- α thermal diffusivity

If $t_{1/2}$ is defined as the time required for the non illuminated surface to experience half its final temperature rise then.

$$\alpha = 0.1388L^2 / t_{1/2}$$

- L is the thickness of the disc

Radioisotope Decay Fuels

Radioisotope Battery

The terms atomic battery, nuclear battery and radioisotope battery are used interchangeably to describe a device which uses the radioactive decay to generate electricity. These systems use radioisotopes that produce low energy beta particles or sometimes alpha particles of varying energies. Low energy beta particles are needed to prevent the production of high energy penetrating bremsstrahlung radiation that would require heavy shielding. Radioisotopes such as plutonium-238, curium-242, curium-244 and strontium-90 have been used. Tritium, nickel-63, promethium-147, and technetium-99 have been tested.

There are two main categories of atomic batteries: thermal and non-thermal. The non-thermal atomic batteries, which have many different designs, exploit charged alpha and beta particles. These designs include the direct charging generators, betavoltaics, the optoelectric nuclear battery, and the radioisotope piezoelectric generator. The thermal atomic batteries on the other hand, convert the heat from the radioactive decay to electricity. These designs include thermionic converter, thermophotovoltaic cells, alkali-metal thermal to electric converter, and the most common design, the radioisotope thermoelectric generator.

Radioisotope Thermoelectric Generators

A *radioisotope thermoelectric generator* (RTG) is a simple electrical generator which converts heat into electricity from a radioisotope using an array of thermocouples.

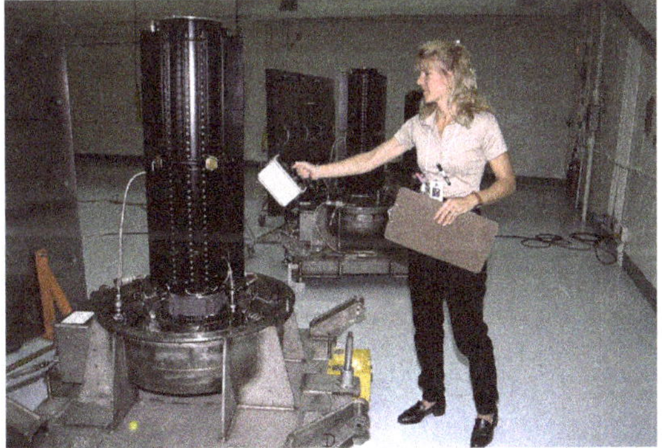

Inspection of Cassini spacecraft RTGs before launch

^{238}Pu has become the most widely used fuel for RTGs, in the form of plutonium dioxide. It has a half-life of 87.7 years, reasonable energy density, and exceptionally low gamma and neutron radiation levels. Some Russian terrestrial RTGs have used ^{90}Sr; this isotope has a shorter half-life and a much lower energy density, but is cheaper. Early RTGs, first built in 1958 by the U.S. Atomic Energy Commission, have used ^{210}Po. This fuel provides phenomenally huge energy density, (a single gram of polonium-210 generates 140 watts thermal) but has limited use because of its very short half-life and gamma production, and has been phased out of use for this application.

Radioisotope Heater Units (RHU)

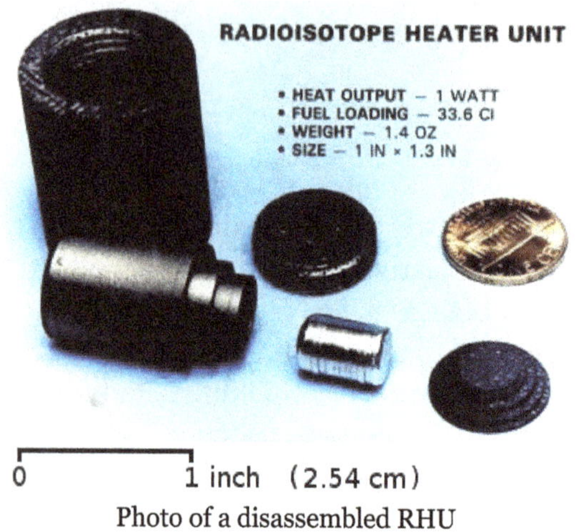

Photo of a disassembled RHU

Radioisotope heater units normally provide about 1 watt of heat each, derived from the decay of a few grams of plutonium-238. This heat is given off continuously for several decades.

Their function is to provide highly localised heating of sensitive equipment (such as electronics in outer space). The Cassini–Huygens orbiter to Saturn contains 82 of these units (in addition to its 3 main RTG's for power generation). The Huygens probe to Titan contains 35 devices.

Fusion Fuels

Fusion fuels include deuterium (^2H) and tritium (^3H) as well as helium-3 (^3He). Many other elements can be fused together, but the larger electrical charge of their nuclei means that much higher temperatures are required. Only the fusion of the lightest elements is seriously considered as a future energy source. Fusion of the lightest atom, ^1H hydrogen, as is done in the Sun and stars, has also not been considered practical on Earth. Although the energy density of fusion fuel is even higher than fission fuel, and fusion reactions sustained for a few minutes have been achieved, utilizing fusion fuel as a net energy source remains only a theoretical possibility.

First-generation Fusion Fuel

Deuterium and tritium are both considered first-generation fusion fuels; they are the easiest to fuse, because the electrical charge on their nuclei is the lowest of all elements. The three most commonly cited nuclear reactions that could be used to generate energy are:

$$^2H + ^3H \rightarrow n(14.07\ MeV) + ^4He\ (3.52\ MeV)$$

$$^2H + ^2H \rightarrow n(2.45\ MeV) + ^3He\ (0.82\ MeV)$$

$$^2H + ^2H \rightarrow p(3.02\ MeV) + ^3H\ (1.01\ MeV)$$

Second-generation Fusion Fuel

Second-generation fuels require either higher confinement temperatures or longer confinement time than those required of first-generation fusion fuels, but generate fewer neutrons. Neutrons are an unwanted byproduct of fusion reactions in an energy generation context, because they are absorbed by the walls of a fusion chamber, making them radioactive. They cannot be confined by magnetic fields, because they are not electrically charged. This group consists of deuterium and helium-3. The products are all charged particles, but there may be significant side reactions leading to the production of neutrons.

$$^{2}H + ^{3}He \rightarrow p(14.68\ MeV) + ^{4}He\ (3.67\ MeV)$$

Third-generation Fusion Fuel

Third-generation fusion fuels produce only charged particles in the primary reactions, and side reactions are relatively unimportant. Since a very small amount of neutrons is produced, there would be little induced radioactivity in the walls of the fusion chamber. This is often seen as the end goal of fusion research. ^{3}He has the highest Maxwellian reactivity of any 3rd generation fusion fuel. However, there are no significant natural sources of this substance on Earth.

$$^{3}He + ^{3}He \rightarrow 2p + ^{4}He\ (12.86\ MeV)$$

Another potential aneutronic fusion reaction is the proton-boron reaction:

$$p + ^{11}B \rightarrow 3\,^{4}He\ (8.7\ MeV)$$

Under reasonable assumptions, side reactions will result in about 0.1% of the fusion power being carried by neutrons. With 123 keV, the optimum temperature for this reaction is nearly ten times higher than that for the pure hydrogen reactions, the energy confinement must be 500 times better than that required for the D-T reaction, and the power density will be 2500 times lower than for D-T.

Fuel Comparison for Nuclear Reactors

The nuclear reactors in operation around the globe utilize uranium (or) plutonium- based materials as fuel. The fuel may be in the form of a metal or metal oxide, or metal carbide or metal nitride.

Let us look at the important properties of fuel and their implications, which are shown in Table. These properties serve as the criteria for selection of fuel materials.

Table: Important thermo-physical properties of nuclear fuels and their implications

Property	Implications
Nuclear properties: Absorption cross section, fission cross section, fission products, neutron production, energy released	Determines the mass of fissile required, volume of the core

Thermal conductivity	Indicates the ability of a material to transfer heat by thermal conduction. Higher values of thermal conductivity are desirable for fuels. This would facilitate rapid removal of heat from the fuel. As a result, with fuel of higher thermal conductivity the temperature gradient in the fuel will be low. In nutshell, higher thermal conductivity fuel permits the reactor operation at high power density and specific power (power per unit mass of fuel) without leading to excessive temperature gradients.
Melting point	Defines the limiting power of a fuel element. Reason: The maximum temperature that a fuel can be allowed to reach is influenced by the melting point. Lower the melting point of the fuel, lower is the maximum permissible temperature and the power of a fuel. Hence, fuels with higher melting points are desired
Dimensional stability	The fuel must exhibit dimensional stability at high temperatures and neutron irradiations. Increased burnup must not result in compromise of dimensional stability. Swelling results in compromise of dimensional stability. Swelling may arise either due to high temperature or neutron irradiation or both. A fuel with higher tendency to swell is likely to constrict the coolant channels. With excessive swelling, fuel may come in contact with cladding resulting in rupture of cladding and release of fission gases. A ceramic fuel like UO2 retains fission gases more effectively and is less prone to swelling.
Coefficient of thermal expansion	A measure of expansion of a material with increase in temperature. It may be recalled that a small gap exists between the fuel pins and the cladding in the radial direction called fuel-cladding gap. When the fuel and cladding have different coefficients of thermal expansion then one of the following scenario may happen: (i) If the thermal expansion coefficient of fuel is more than that of cladding, the fuel expands more than that of cladding during heating resulting in reduced fuel-cladding gap. Eventually fuel and cladding may come into contact with each other. (ii) If the thermal expansion coefficient of cladding is more than that of fuel, cladding expands more than that of the fuel during heating resulting in increased fuel-cladding gap. The fuel-cladding gap contains (fission) gases that are poor conductors of heat. With increased fuel-cladding gap, the total thermal resistance for heat transfer between fuel and coolant increases. This causes increase in the temperature of the fuel due to reduced heat transfer.
Chemical properties	In case of cladding failure in terms of cracks, there will be direct contact between fuel and coolant.
	Under these circumstances, chemical stability of fuel while in contact with coolant is important.

Oxide Fuels

The important examples in the category of oxide fuels are UO_2 (Uranium dioxide) or mixed oxide (UO_2-PuO_2). The major disadvantages of oxide fuels are their lower density and lower thermal conductivity. Lower density leads to the requirement of a larger core diameter. One may be aware of the fact that to pack a certain mass of material of lower density, a relatively large volume of space is required. Since the density of oxide is low, their use requires reactors of larger diameter to be fabricated.

Lower thermal conductivity leads to poor transfer of heat from the centre of the fuel pin to the edge of the pin and to the cladding. Hence the temperature at the centre of the fuel must be sufficiently high to ensure that the outer surface is sufficiently hot to transfer heat to the coolant. However, the fuel temperature cannot be allowed to increase beyond certain limits. If the maximum allowable temperature of the fuel is T_{max} and that of the cladding is T_{clad}, the lower thermal conductivity of oxide leads to a large difference between T_{max} and T_{clad}. Since the maximum allowable temperature of the fuel limits T_{max}, limitation exists on T_{clad} as well. This puts a limitation on the maximum temperature of primary coolant and hence that of the steam generated.

The comparison of temperature profiles with two different types of fuels: (a) lower thermal conductivity fuel and (b) higher thermal conductivity fuel. These profiles are shown for a fixed fuel centre-line temperature. The temperature of a fuel is maximum at the centre-line and decreases radially outward in a cylindrical fuel element. A temperature difference exists in the fuel-clad gap due to higher resistance in this gap occupied by fission gases. A small resistance exists in the cladding layer as well. All these result in difference between the maximum temperature of the fuel (T_{max}) and the cladding surface temperature (T_c) with which the coolant is in contact.

For a constant heat transfer rate, a higher temperature gradient (dT/dr) indicates higher resistance to heat transfer. The gradient may be split up into two parts: (i) temperature gradient in the fuel and (ii) temperature gradient in fuel-clad gap and in clad material. While comparing two fuels of different thermal conductivities, the temperature gradient in the fuel is more whose thermal conductivity is lower, when compared to that in a fuel of higher thermal conductivity. Note that the temperature gradient in fuel-clad gap and in clad is same in both Figures a and b. However, the temperature gradient is high due to use of low thermal conductivity fuel.

Nuclear Fuel Cycle

The nuclear fuel cycle, also called nuclear fuel chain, is the progression of nuclear fuel through a series of differing stages. It consists of steps in the *front end*, which are the preparation of the fuel, steps in the *service period* in which the fuel is used during reactor operation, and steps in the *back end*, which are necessary to safely manage, contain, and either reprocess or dispose of spent nuclear fuel. If spent fuel is not reprocessed, the fuel cycle is referred to as an *open fuel cycle* (or a *once-through fuel cycle*); if the spent fuel is reprocessed, it is referred to as a *closed fuel cycle*.

Basic Concepts

Nuclear power relies on fissionable material that can sustain a chain reaction with neutrons. Examples of such materials include Uranium and Plutonium. Most nuclear reactors use a moderator to lower the kinetic energy of the neutrons and increase the probability that fission will occur. This allows reactors to use material with far lower concentration of fissile isotopes than are needed for nuclear weapons. Graphite and heavy water are the most effective moderators, because they slow the neutrons through collisions without absorbing them. Reactors using heavy water or graphite as the moderator can operate using natural uranium.

A light water reactor (LWR) uses water in the form that occurs in nature, and requires fuel enriched to higher concentrations of fissile isotopes. Typically LWRs use uranium enriched to 3–5% content of the less common isotope U-235, the only fissile isotope that is found in significant quantity in nature. One alternative to this low-enriched uranium (LEU) fuel are mixed oxide (MOX) fuels produced by blending plutonium with natural or depleted uranium, and these fuels provide an avenue to utilize surplus weapons-grade plutonium. Another type of MOX fuel involves mixing LEU with thorium, which generates the fissile isotope U-233. Both plutonium and U-233 are produced from the absorption of neutrons by irradiating fertile materials in a reactor, in particular the common uranium isotope U-238 and thorium, respectively, and can be separated from spent uranium and thorium fuels in reprocessing plants.

Some reactors do not use moderators to slow the neutrons. Like nuclear weapons, which also use unmoderated or "fast" neutrons, these fast-neutron reactors require much higher concentrations of fissile isotopes in order to sustain a chain reaction. They are also capable of breeding fissile isotopes from fertile materials; a breeder reactor is one that generates more fissile material in this way than it consumes.

During the nuclear reaction inside a reactor, the fissile isotopes in nuclear fuel are consumed, producing more and more fission products, most of which are considered radioactive waste. The buildup of fission products and consumption of fissile isotopes eventually stop the nuclear reaction, causing the fuel to become a spent nuclear fuel. When 3% enriched LEU fuel is used, the spent fuel typically consists of roughly 1% U-235, 95% U-238, 1% plutonium and 3% fission products. Spent fuel and other high-level radioactive waste is extremely hazardous, although nuclear reactors produce relatively small volumes of waste compared to other power plants because of the high energy density of nuclear fuel. Safe management of these byproducts of nuclear power, including their storage and disposal, is a difficult problem for any country using nuclear power.

Front End

Uranium ore - the principal raw material of nuclear fuel

Yellowcake - the form in which uranium is transported to a conversion plant

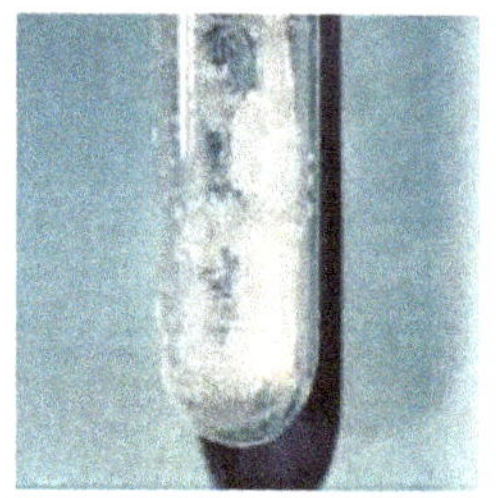

UF_6 - used in enrichmen

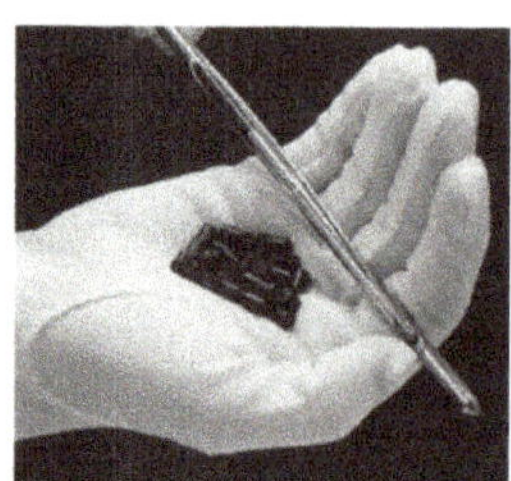

Nuclear fuel - a compact, inert, insoluble solid

Exploration

A deposit of uranium, such as uraninite, discovered by geophysical techniques, is evaluated and sampled to determine the amounts of uranium materials that are extractable at specified costs from the deposit. Uranium reserves are the amounts of ore that are estimated to be recoverable at stated costs.

Naturally occurring uranium consists primarily of two isotopes U-238 and U-235, with 99.28% of the metal being U-238 while 0.71% is U-235, and the remaining 0.01% is mostly U-234. The number in such names refers to the isotope's atomic mass number, which is the number of protons plus the number of neutrons in the atomic nucleus.

The atomic nucleus of U-235 will nearly always fission when struck by a free neutron, and the isotope is therefore said to be a "fissile" isotope. The nucleus of a U-238 atom on the other hand, rather than undergoing fission when struck by a free neutron, will nearly always absorb the neutron and yield an atom of the isotope U-239. This isotope then undergoes natural radioactive decay to yield Pu-239, which, like U-235, is a fissile isotope. The atoms of U-238 are said to be fertile, because, through neutron irradiation in the core, some eventually yield atoms of fissile Pu-239.

Mining

Uranium ore can be extracted through conventional mining in open pit and underground methods similar to those used for mining other metals. In-situ leach mining methods also are used to mine uranium in the United States. In this technology, uranium is leached from the in-place ore through an array of regularly spaced wells and is then recovered from the leach solution at a surface plant. Uranium ores in the United States typically range from about 0.05 to 0.3% uranium oxide (U_3O_8). Some uranium deposits developed in other countries are of higher grade and are also larger than deposits mined in the United States. Uranium is also present in very low-grade amounts (50 to 200 parts per million) in some domestic phosphate-bearing deposits of marine origin. Because very large quantities of phosphate-bearing rock are mined for the production of wet-process phosphoric acid used in high analysis fertilizers and other phosphate chemicals, at some phosphate processing plants the uranium, although present in very low concentrations, can be economically recovered from the process stream.

Milling

Mined uranium ores normally are processed by grinding the ore materials to a uniform particle size and then treating the ore to extract the uranium by chemical leaching. The milling process commonly yields dry powder-form material consisting of natural uranium, "yellowcake", which is sold on the uranium market as U_3O_8. Note that the material isnt allways yellow.

Uranium Conversion

Usually milled Uranium oxide, U_3O_8 (Triuranium octaoxide) is then processed into either of 2 substances depending on the intended use.

For use in most reactors U_3O_8 is usually converted to Uranium hexafluoride (UF_6), the input stock for most commercial uranium enrichment facilities. A solid at room temperature, uranium hexafluoride becomes gaseous at 57 °C (134 °F). At this stage of the cycle the uranium hexafluoride conversion product still has the natural isotopic mix (99.28% of U-238 plus 0.71% of U-235).

For used in reactors such as CANDU which do not require enriched fuel, the U_3O_8 may instead be converted to uranium dioxide (UO_2) which can be included in ceramic fuel elements.

In the current nuclear industry the volume of material converted directly to UO_2 is typically quite small compared to that converted to UF_6.

Enrichment

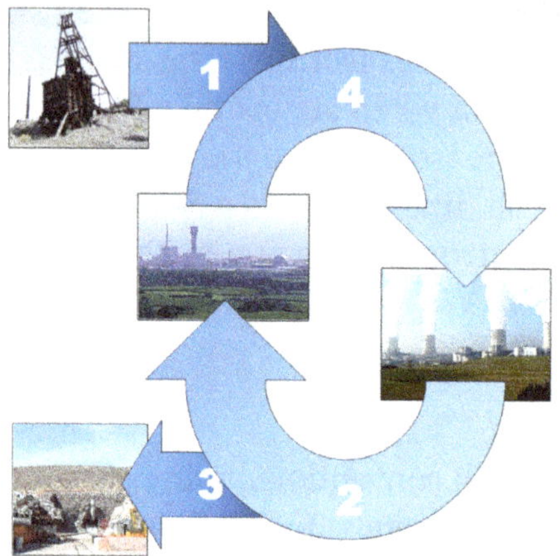

Nuclear fuel cycle begins when uranium is mined, enriched and manufactured to nuclear fuel (1) which is delivered to a nuclear power plant. After usage in the power plant the spent fuel is delivered to a reprocessing plant (if fuel is recycled) (2) or to a final repository (if no recycling is done) (3) for geological disposition. In reprocessing 95% of spent fuel can be recycled to be returned to usage in a nuclear power plant (4).

The natural concentration (0.71%) of the fissionable isotope U-235 is less than that required to sustain a nuclear chain reaction in light water reactor cores. Accordingly UF_6 produced from natural Uranium sources must be enriched to a higher concentration of the fissionable isotope before being used as nuclear fuel in such reactors. The level of enrichment for a particular nuclear fuel order is specified by the customer according to the application they will use it for: light-water reactor fuel normally is enriched to 3.5% U-235, but uranium enriched to lower concentrations is also required. Enrichment is accomplished using any of several methods of isotope separation. Gaseous diffusion and gas centrifuge are the commonly used uranium enrichment methods, but new enrichment technologies are currently being developed.

The bulk (96%) of the byproduct from enrichment is depleted uranium (DU), which can be used for armor, kinetic energy penetrators, radiation shielding and ballast. As of 2008 there are vast quantities of depleted uranium in storage. The United States Department of Energy alone has 470,000 tonnes. About 95% of depleted uranium is stored as uranium hexafluoride (UF_6).

Fabrication

For use as nuclear fuel, enriched uranium hexafluoride is converted into uranium dioxide (UO_2) powder that is then processed into pellet form. The pellets are then fired in a high temperature sintering furnace to create hard, ceramic pellets of enriched uranium. The cylindrical pellets then undergo a grinding process to achieve a uniform pellet size. The pellets are stacked, according to each nuclear reactor core's design specifications, into tubes of corrosion-resistant metal alloy. The tubes are sealed to contain the fuel pellets: these tubes are called fuel rods. The finished fuel rods are grouped in special fuel assemblies that are then used to build up the nuclear fuel core of a power reactor.

The alloy used for the tubes depends on the design of the reactor. Stainless steel was used in the past, but most reactors now use a zirconium alloy. For the most common types of reactors, boiling water reactors (BWR) and pressurized water reactors (PWR), the tubes are assembled into bundles with the tubes spaced precise distances apart. These bundles are then given a unique identification number, which enables them to be tracked from manufacture through use and into disposal.

Service Period

Transport of Radioactive Materials

Transport is an integral part of the nuclear fuel cycle. There are nuclear power reactors in operation in several countries but uranium mining is viable in only a few areas. Also, in the course of over forty years of operation by the nuclear industry, a number of specialized facilities have been developed in various locations around the world to provide fuel cycle services and there is a need to transport nuclear materials to and from these facilities. Most transports of nuclear fuel material occur between different stages of the cycle, but occasionally a material may be transported between similar facilities. With some exceptions, nuclear fuel cycle materials are transported in solid form, the exception being uranium hexafluoride (UF_6) which is considered a gas. Most of the material used in nuclear fuel is transported several times during the cycle. Transports are frequently international, and are often over large distances. Nuclear materials are generally transported by specialized transport companies.

Since nuclear materials are radioactive, it is important to ensure that radiation exposure of those involved in the transport of such materials and of the general public along transport routes is limited. Packaging for nuclear materials includes, where appropriate, shielding to reduce potential radiation exposures. In the case of some materials, such as fresh uranium fuel assemblies, the radiation levels are negligible and no shielding is required. Other materials, such as spent fuel and high-level waste, are highly radioactive and require special handling. To limit the risk in transporting highly radioactive materials, containers known as spent nuclear fuel shipping casks are used which are designed to maintain integrity under normal transportation conditions and during hypothetical accident conditions.

In-core Fuel Management

A nuclear reactor core is composed of a few hundred "assemblies", arranged in a regular array of cells, each cell being formed by a fuel or control rod surrounded, in most designs, by a moderator and coolant, which is water in most reactors.

Because of the fission process that consumes the fuels, the old fuel rods must be replaced periodically with fresh ones (this is called a (replacement) cycle). During a given replacement cycle only some of the assemblies (typically one-third) are replaced since fuel depletion occurs at different rates at different places within the reactor core. Furthermore, for efficiency reasons, it is not a good policy to put the new assemblies exactly at the location of the removed ones. Even bundles of the same age will have different burn-up levels due to their previous positions in the core. Thus the available bundles must be arranged in such a way that the yield is maximized, while safety limitations and operational constraints are satisfied. Consequently, reactor operators are faced with the so-called optimal fuel reloading problem, which consists of optimizing the rearrangement of all the assemblies, the old and fresh ones, while still maximizing the reactivity of the reactor core so as to maximise fuel burn-up and minimise fuel-cycle costs.

This is a discrete optimization problem, and computationally infeasible by current combinatorial methods, due to the huge number of permutations and the complexity of each computation. Many numerical methods have been proposed for solving it and many commercial software packages have been written to support fuel management. This is an ongoing issue in reactor operations as no definitive solution to this problem has been found. Operators use a combination of computational and empirical techniques to manage this problem.

The Study of Used Fuel

Used nuclear fuel is studied in Post irradiation examination, where used fuel is examined to know more about the processes that occur in fuel during use, and how these might alter the outcome of an accident. For example, during normal use, the fuel expands due to thermal expansion, which can cause cracking. Most nuclear fuel is uranium dioxide, which is a cubic solid with a structure similar to that of calcium fluoride. In used fuel the solid state structure of most of the solid remains the same as that of pure cubic uranium dioxide. SIMFUEL is the name given to the simulated spent fuel which is made by mixing finely ground metal oxides, grinding as a slurry, spray drying it before heating in hydrogen/argon to 1700 °C. In SIMFUEL, 4.1% of the volume of the solid was in the form of metal nanoparticles which are made of molybdenum, ruthenium, rhodium and palladium. Most of these metal particles are of the ε phase (hexagonal) of Mo-Ru-Rh-Pd alloy, while smaller amounts of the α (cubic) and σ (tetragonal) phases of these metals were found in the SIMFUEL. Also present within the SIMFUEL was a cubic perovskite phase which is a barium strontium zirconate ($Ba_xSr_{1-x}ZrO_3$).

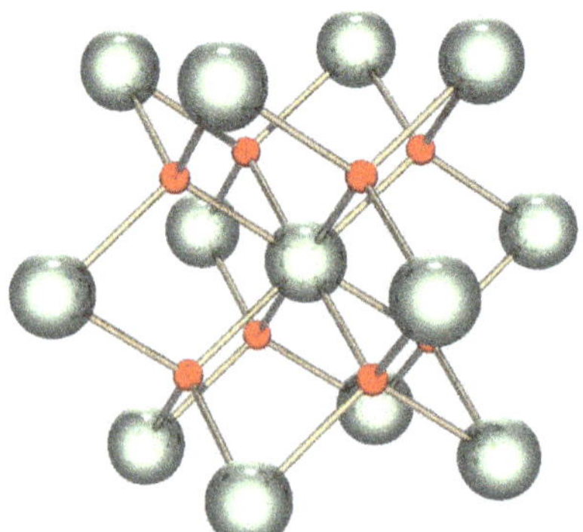

The solid state structure of uranium dioxide, the oxygen atoms are in green and the uranium atoms in red

Uranium dioxide is very insoluble in water, but after oxidation it can be converted to uranium trioxide or another uranium(VI) compound which is much more soluble. Uranium dioxide (UO_2) can be oxidised to an oxygen rich hyperstoichiometric oxide (UO_{2+x}) which can be further oxidised to U_4O_9, U_3O_7, U_3O_8 and $UO_3.2H_2O$.

Because used fuel contains alpha emitters (plutonium and the minor actinides), the effect of adding an alpha emitter (^{238}Pu) to uranium dioxide on the leaching rate of the oxide has been investigated. For the crushed oxide, adding ^{238}Pu tended to increase the rate of leaching, but the difference in the leaching rate between 0.1 and 10% ^{238}Pu was very small.

The concentration of carbonate in the water which is in contact with the used fuel has a considerable effect on the rate of corrosion, because uranium(VI) forms soluble anionic carbonate complexes such as $[UO_2(CO_3)_2]^{2-}$ and $[UO_2(CO_3)_3]^{4-}$. When carbonate ions are absent, and the water is not strongly acidic, the hexavalent uranium compounds which form on oxidation of uranium dioxide often form insoluble hydrated uranium trioxide phases.

Thin films of uranium dioxide can be deposited upon gold surfaces by 'sputtering' using uranium metal and an argon/oxygen gas mixture, . These gold surfaces modified with uranium dioxide have been used for both cyclic voltammetry and AC impedance experiments, and these offer an insight into the likely leaching behaviour of uranium dioxide.

Fuel Cladding Interactions

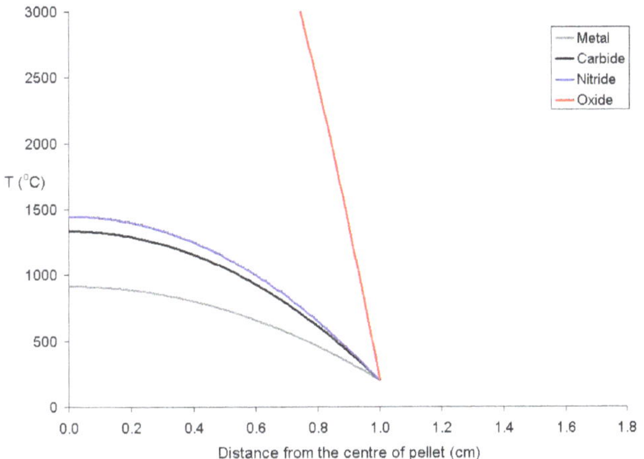

Temperature profile for a 20 mm diameter fuel pellet with a power density of 1 kW per cubic meter. The fuels other than uranium dioxide are not compromised.

The study of the nuclear fuel cycle includes the study of the behaviour of nuclear materials both under normal conditions and under accident conditions. For example, there has been much work on how uranium dioxide based fuel interacts with the zirconium alloy tubing used to cover it. During use, the fuel swells due to thermal expansion and then starts to react with the surface of the zirconium alloy, forming a new layer which contains both fuel and zirconium (from the cladding). Then, on the fuel side of this mixed layer, there is a layer of fuel which has a higher caesium to uranium ratio than most of the fuel. This is because xenon isotopes are formed as fission products that diffuse out of the lattice of the fuel into voids such as the narrow gap between the fuel and the cladding. After diffusing into these voids, it decays to caesium isotopes. Because of the thermal gradient which exists in the fuel during use, the volatile fission products tend to be driven from the centre of the pellet to the rim area. Below is a graph of the temperature of uranium metal, uranium nitride and uranium dioxide as a function of distance from the centre of a 20 mm diameter pellet with a rim temperature of 200 °C. The uranium dioxide (because of its poor thermal conductivity) will overheat at the centre of the pellet, while the other more thermally conductive forms of uranium remain below their melting points.

Normal and Abnormal Conditions

The nuclear chemistry associated with the nuclear fuel cycle can be divided into two main areas; one area is concerned with operation under the intended conditions while the other area is concerned with maloperation conditions where some alteration from the normal operating conditions has occurred or (*more rarely*) an accident is occurring.

The releases of radioactivity from normal operations are the small planned releases from uranium ore processing, enrichment, power reactors, reprocessing plants and waste stores. These can be in different chemical/physical form from releases which could occur under accident conditions. In addition the isotope signature of a hypothetical accident may be very different from that of a planned normal operational discharge of radioactivity to the environment.

Just because a radioisotope is released it does not mean it will enter a human and then cause harm. For instance, the migration of radioactivity can be altered by the binding of the radioisotope to the surfaces of soil particles. For example, caesium (Cs) binds tightly to clay minerals such as illite and montmorillonite, hence it remains in the upper layers of soil where it can be accessed by plants with shallow roots (such as grass). Hence grass and mushrooms can carry a considerable amount of ^{137}Cs which can be transferred to humans through the food chain. But ^{137}Cs is not able to migrate quickly through most soils and thus is unlikely to contaminate well water. Colloids of soil minerals can migrate through soil so simple binding of a metal to the surfaces of soil particles does not completely fix the metal.

According to Jiří Hála's text book, the distribution coefficient K_d is the ratio of the soil's radioactivity (Bq g^{-1}) to that of the soil water (Bq ml^{-1}). If the radioisotope is tightly bound to the minerals in the soil, then less radioactivity can be absorbed by crops and grass growing on the soil.

- Cs-137 K_d = 1000
- Pu-239 K_d = 10000 to 100000
- Sr-90 K_d = 80 to 150
- I-131 K_d = 0.007 to 50

In dairy farming one of the best countermeasures against ^{137}Cs is to mix up the soil by deeply ploughing the soil. This has the effect of putting the ^{137}Cs out of reach of the shallow roots of the grass, hence the level of radioactivity in the grass will be lowered. Also after a nuclear war or serious accident, the removal of top few cm of soil and its burial in a shallow trench will reduce the long-term gamma dose to humans due to ^{137}Cs, as the gamma photons will be attenuated by their passage through the soil.

Even after the radioactive element arrives at the roots of the plant, the metal may be rejected by the biochemistry of the plant. The details of the uptake of ^{90}Sr and ^{137}Cs into sunflowers grown under hydroponic conditions has been reported. The caesium was found in the leaf veins, in the stem and in the apical leaves. It was found that 12% of the caesium entered the plant, and 20% of the strontium. This paper also reports details of the effect of potassium, ammonium and calcium ions on the uptake of the radioisotopes.

In livestock farming, an important countermeasure against ^{137}Cs is to feed animals a small amount of Prussian blue. This iron potassium cyanide compound acts as an ion-exchanger. The cyanide is so tightly bonded to the iron that it is safe for a human to eat several grams of Prussian blue per day. The Prussian blue reduces the biological half-life (different from the nuclear half-life) of the caesium. The physical or nuclear half-life of ^{137}Cs is about 30 years. This is a constant which can not be changed but the biological half-life is not a constant. It will change according to the nature and habits of the organism for which it is expressed. Caesium in humans normally has a biological half-life of between one and four months. An added advantage of the Prussian blue is that the caesium which is stripped from the animal in the droppings is in a form which is not available to plants. Hence it prevents the caesium from being recycled. The form of Prussian blue required for the treatment of humans or animals is a special grade. Attempts to use the pigment grade used in paints have not been successful. Note that a good source of data on the subject of caesium in Chernobyl fallout exists at (*Ukrainian Research Institute for Agricultural Radiology*).

Release of Radioactivity from Fuel During Normal use and Accidents

The IAEA assume that under normal operation the coolant of a water-cooled reactor will contain some radioactivity but during a reactor accident the coolant radioactivity level may rise. The IAEA states that under a series of different conditions different amounts of the core inventory can be released from the fuel, the four conditions the IAEA consider are *normal operation*, a spike in coolant activity due to a sudden shutdown/loss of pressure (core remains covered with water), a cladding failure resulting in the release of the activity in the fuel/cladding gap (this could be due to the fuel being uncovered by the loss of water for 15–30 minutes where the cladding reached a temperature of 650-1250 °C) or a melting of the core (the fuel will have to be uncovered for at least 30 minutes, and the cladding would reach a temperature in excess of 1650 °C).

Based upon the assumption that a Pressurized water reactor contains 300 tons of water, and that the activity of the fuel of a 1 GWe reactor is as the IAEA predicts, then the coolant activity after an accident such as the Three Mile Island accident (where a core is uncovered and then recovered with water) can be predicted.

Releases from Reprocessing under Normal Conditions

It is normal to allow used fuel to stand after the irradiation to allow the short-lived and radiotoxic iodine isotopes to decay away. In one experiment in the USA, fresh fuel which had not been allowed to decay was reprocessed (the Green run) to investigate the effects of a large iodine release from the reprocessing of short cooled fuel. It is normal in reprocessing plants to scrub the off gases from the dissolver to prevent the emission of iodine. In addition to the emission of iodine the noble gases and tritium are released from the fuel when it is dissolved. It has been proposed that by voloxidation (heating the fuel in a furnace under oxidizing conditions) the majority of the tritium can be recovered from the fuel.

A paper was written on the radioactivity in oysters found in the Irish Sea. These were found by gamma spectroscopy to contain ^{141}Ce, ^{144}Ce, ^{103}Ru, ^{106}Ru, ^{137}Cs, ^{95}Zr and ^{95}Nb. Additionally, a zinc activation product (^{65}Zn) was found, which is thought to be due to the corrosion of magnox fuel cladding in spent fuel pools. It is likely that the modern releases of all these isotopes from the Windscale event is smaller.

On-load Reactors

Some reactor designs, such as RBMKs or CANDU reactors, can be refueled without being shut down.

This is achieved through the use of many small pressure tubes to contain the fuel and coolant, as opposed to one large pressure vessel as in pressurized water reactor (PWR) or boiling water reactor (BWR) designs. Each tube can be individually isolated and refueled by an operator-controlled fueling machine, typically at a rate of up to 8 channels per day out of roughly 400 in CANDU reactors. On-load refueling allows for the optimal fuel reloading problem to be dealt with continuously, leading to more efficient use of fuel. This increase in efficiency is partially offset by the added complexity of having hundreds of pressure tubes and the fueling machines to service them.

Interim Storage

After its operating cycle, the reactor is shut down for refueling. The fuel discharged at that time (spent fuel) is stored either at the reactor site (commonly in a spent fuel pool) or potentially in a common facility away from reactor sites. If on-site pool storage capacity is exceeded, it may be desirable to store the now cooled aged fuel in modular dry storage facilities known as Independent Spent Fuel Storage Installations (ISFSI) at the reactor site or at a facility away from the site. The spent fuel rods are usually stored in water or boric acid, which provides both cooling (the spent fuel continues to generate decay heat as a result of residual radioactive decay) and shielding to protect the environment from residual ionizing radiation, although after at least a year of cooling they may be moved to dry cask storage.

Reprocessing

The Sellafield reprocessing plant

Spent fuel discharged from reactors contains appreciable quantities of fissile (U-235 and Pu-239), fertile (U-238), and other radioactive materials, including reaction poisons, which is why the fuel had to be removed. These fissile and fertile materials can be chemically separated and recovered from the spent fuel. The recovered uranium and plutonium can, if economic and institutional conditions permit, be recycled for use as nuclear fuel. This is currently not done for civilian spent nuclear fuel in the United States.

Mixed oxide, or MOX fuel, is a blend of reprocessed uranium and plutonium and depleted uranium which behaves similarly, although not identically, to the enriched uranium feed for which most nuclear reactors were designed. MOX fuel is an alternative to low-enriched uranium (LEU) fuel used in the light water reactors which predominate nuclear power generation.

Currently, plants in Europe are reprocessing spent fuel from utilities in Europe and Japan. Reprocessing of spent commercial-reactor nuclear fuel is currently not permitted in the United States due to the perceived danger of nuclear proliferation. However the recently announced Global Nuclear Energy Partnership would see the U.S. form an international partnership to see spent nuclear fuel reprocessed in a way that renders the plutonium in it usable for nuclear fuel but not for nuclear weapons.

Partitioning and Transmutation

As an alternative to the disposal of the PUREX raffinate in glass or Synroc, the most radiotoxic elements can be removed through advanced reprocessing. After separation, the minor actinides and some long-lived fission products can be converted to short-lived isotopes by either neutron or photon irradiation. This is called transmutation.

Waste Disposal

Actinides and fission products by half-life

Actinides by decay chain				**Half-life range (y)**	**Fission products of ^{235}U by yield**			
$4n$	$4n$+1	$4n$+2	$4n$+3			4.5–7%	0.04–1.25%	<0.001%
^{228}Ra№				**4–6**			^{155}Euþ	
^{244}Cm^{f}	^{241}Pu^{f}	^{250}Cf	^{227}Ac№	**10–29**	†	^{90}Sr	^{85}Kr	^{113m}Cdþ
^{232}U^{f}		^{238}Pu^{f}№	^{243}Cm^{f}	**29–97**		^{137}Cs	^{151}Smþ	^{121m}Sn
^{248}Bk	^{249}Cf^{f}	^{242m}Am^{f}		**141–351**	No fission products have a half-life in the range of 100–210 k years ...			
	^{241}Am^{f}		^{251}Cf^{f}	**430–900**				
		^{226}Ra№	^{247}Bk	**1.3 k – 1.6 k**				
^{240}Pu^{f}№	^{229}Th№	^{246}Cm^{f}	^{243}Am^{f}	**4.7 k – 7.4 k**				
	^{245}Cm^{f}	^{250}Cm		**8.3 k – 8.5 k**				
			^{239}Pu^{f}№	**24.1 k**				
		^{230}Th№	^{231}Pa№	**32 k – 76 k**				
^{236}Np^{f}	^{233}U^{f}№	^{234}U№		**150 k – 250 k**	‡	^{99}Tc¢	^{126}Sn	
^{248}Cm		^{242}Pu^{f}		**327 k – 375 k**			^{79}Se¢	
				1.53 M		^{93}Zr		
	^{237}Np^{f}№			**2.1 M – 6.5 M**		^{135}Cs¢	^{107}Pd	
^{236}U№			^{247}Cm^{f}	**15 M – 24 M**			^{129}I¢	
^{244}Pu№				**80 M**	... nor beyond 15.7 M years			
^{232}Th№		^{238}U№	^{235}U^{f}№	**0.7 G – 14.1 G**				

Legend for superscript symbols
¢ has thermal neutron capture cross section in the range of 8–50 barns
f fissile
m metastable isomer
№ naturally occurring radioactive material (NORM)
þ neutron poison (thermal neutron capture cross section greater than 3k barns)
† range 4–97 y: Medium-lived fission product
‡ over 200,000 y: Long-lived fission product

A current concern in the nuclear power field is the safe disposal and isolation of either spent fuel from reactors or, if the reprocessing option is used, wastes from reprocessing plants. These materials must be isolated from the biosphere until the radioactivity contained in them has diminished to a safe level. In the U.S., under the Nuclear Waste Policy Act of 1982 as amended, the Department of Energy has responsibility for the development of the waste disposal system for spent nuclear fuel and high-level radioactive waste. Current plans call for the ultimate disposal of the wastes in solid form in a licensed deep, stable geologic structure called a deep geological repository. The Department of Energy chose Yucca Mountain as the location for the repository. However, its opening has been repeatedly delayed. Since 1999 thousands of nuclear waste shipments have been stored at the Waste Isolation Pilot Plant in New Mexico.

Fast-neutron reactors can fission all actinides, while the thorium fuel cycle produces low levels of transuranics. Unlike LWRs, in principle these fuel cycles could recycle their plutonium and minor actinides and leave only fission products and activation products as waste. The highly radioactive medium-lived fission products Cs-137 and Sr-90 diminish by a factor of 10 each century; while the long-lived fission products have relatively low radioactivity, often compared favorably to that of the original uranium ore.

Fuel Cycles

Although the most common terminology is *fuel cycle,* some argue that the term *fuel chain* is more accurate, because the spent fuel is never fully recycled. Spent fuel includes fission products, which generally must be treated as waste, as well as uranium, plutonium, and other transuranic elements. Where plutonium is recycled, it is normally reused once in light water reactors, although fast reactors could lead to more complete recycling of plutonium.

Once-through Nuclear Fuel Cycle

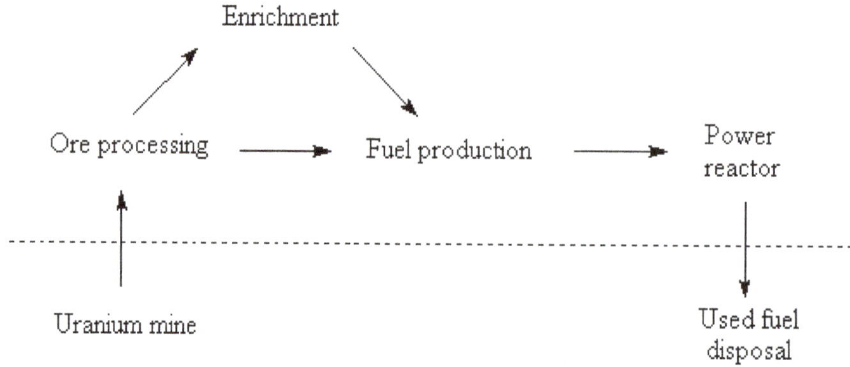

A once through (or open) fuel cycle

Not a cycle *per se*, fuel is used once and then sent to storage without further processing save additional packaging to provide for better isolation from the biosphere. This method is favored by six countries: the United States, Canada, Sweden, Finland, Spain and South Africa. Some countries, notably Finland, Sweden and Canada, have designed repositories to permit future recovery of the material should the need arise, while others plan for permanent sequestration in a geological repository like the Yucca Mountain nuclear waste repository in the United States.

Plutonium Cycle

Several countries, including Japan, Switzerland, and previously Spain and Germany, are using or have used the reprocessing services offered by BNFL and COGEMA. Here, the fission products, minor actinides, activation products, and reprocessed uranium are separated from the reactor-grade plutonium, which can then be fabricated into MOX fuel. Because the proportion of the non-fissile even-mass isotopes of plutonium rises with each pass through the cycle, there are currently no plans to reuse plutonium from used MOX fuel for a third pass in a thermal reactor. However, if fast reactors become available, they may be able to burn these, or almost any other actinide isotopes.

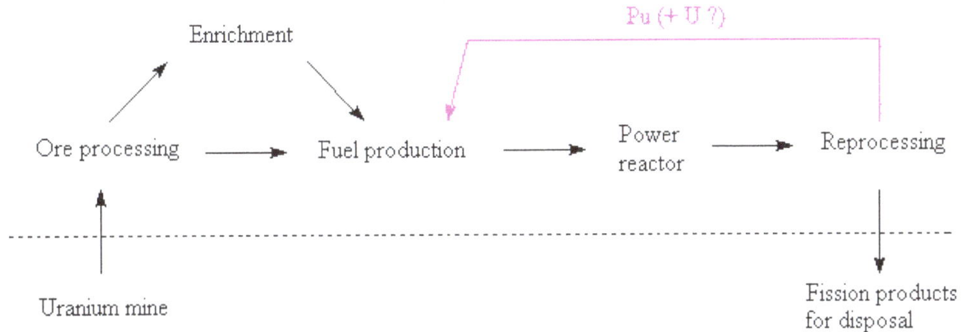

A fuel cycle in which plutonium is used for fuel

Minor Actinides Recycling

It has been proposed that in addition to the use of plutonium, the minor actinides could be used in a critical power reactor. Tests are already being conducted in which americium is being used as a fuel.

A number of reactor designs, like the Integral Fast Reactor, have been designed for this rather different fuel cycle. In principle, it should be possible to derive energy from the fission of any actinide nucleus. With a careful reactor design, all the actinides in the fuel can be consumed, leaving only lighter elements with short half-lives. Whereas this has been done in prototype plants, no such reactor has ever been operated on a large scale.

It so happens that the neutron cross-section of many actinides decreases with increasing neutron energy, but the ratio of fission to simple activation (neutron capture) changes in favour of fission as the neutron energy increases. Thus with a sufficiently high neutron energy, it should be possible to destroy even curium without the generation of the transcurium metals. This could be very desirable as it would make it significantly easier to reprocess and handle the actinide fuel.

One promising alternative from this perspective is an accelerator-driven sub-critical reactor / subcritical reactor. Here a beam of either protons (United States and European designs) or electrons (Japanese design) is directed into a target. In the case of protons, very fast neutrons will spall off the target, while in the case of the electrons, very high energy photons will be generated. These high-energy neutrons and photons will then be able to cause the fission of the heavy actinides.

Such reactors compare very well to other neutron sources in terms of neutron energy:

- Thermal 0 to 100 eV

- Epithermal 100 eV to 100 keV
- Fast (from nuclear fission) 100 keV to 3 MeV
- DD fusion 2.5 MeV
- DT fusion 14 MeV
- Accelerator driven core 200 MeV (lead driven by 1.6 GeV protons)
- Muon-catalyzed fusion 7 GeV.

As an alternative, the curium-244, with a half-life of 18 years, could be left to decay into plutonium-240 before being used in fuel in a fast reactor.

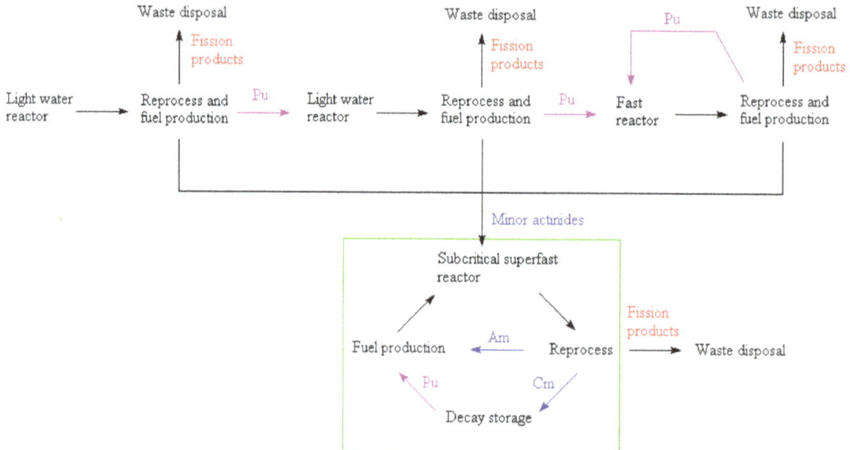

A pair of fuel cycles in which uranium and plutonium are kept separate from the minor actinides. The minor actinide cycle is kept within the green box.

Fuel or Targets for This Actinide Transmutation

To date the nature of the fuel (targets) for actinide transformation has not been chosen.

If actinides are transmuted in a Subcritical reactor it is likely that the fuel will have to be able to tolerate more thermal cycles than conventional fuel. An accelerator-driven sub-critical reactor is unlikely to be able to maintain a constant operation period for equally long times as a critical reactor, and each time the accelerator stops then the fuel will cool down.

On the other hand, if actinides are destroyed using a fast reactor, such as an Integral Fast Reactor, then the fuel will most likely not be exposed to many more thermal cycles than in a normal power station.

Depending on the matrix the process can generate more transuranics from the matrix. This could either be viewed as good (generate more fuel) or can be viewed as bad (generation of more *radiotoxic* transuranic elements). A series of different matrices exists which can control this production of heavy actinides.

Fissile nuclei, like Uranium-235, Plutonium-239 and Uranium-233 respond well to delayed neutrons and are thus important to keep a critical reactor stable, and this limits the amount of minor actinides that can be destroyed in a critical reactor. As a consequence it is important that the chosen matrix allows the reactor to keep the ratio of fissile to non-fissile nuclei high, as this enables it to destroy the long-lived actinides safely. In contrast, the power output of a sub-critical reactor is limited by the intensity of the driving particle accelerator, and thus it need not contain any uranium or plutonium at all. In such a system it may be preferable to have an inert matrix that doesn't produce additional long-lived isotopes.

Actinides in an Inert Matrix

The actinides will be mixed with a metal which will not form more actinides, for instance an alloy of actinides in a solid such as zirconia could be used.

Actinides in a Thorium Matrix

Thorium will on neutron bombardment form uranium-233. U-233 is fissile, and has a larger fission cross section than both U-235 and U-238, and thus it is far less likely to produce higher actinides through neutron capture.

Actinides in a Uranium Matrix

If the actinides are incorporated into a uranium-metal or uranium-oxide matrix, then the neutron capture of U-238 is likely to generate new plutonium-239. An advantage of mixing the actinides with uranium and plutonium is that the large fission cross sections of U-235 and Pu-239 for the less energetic delayed-neutrons could make the reaction stable enough to be carried out in a critical fast reactor, which is likely to be both cheaper and simpler than an accelerator driven system.

Mixed Matrix

It is also possible to create a matrix made from a mix of the above-mentioned materials. This is most commonly done in fast reactors where one may wish to keep the breeding ratio of new fuel high enough to keep powering the reactor, but still low enough that the generated actinides can be safely destroyed without transporting them to another site. One way to do this is to use fuel where actinides and uranium is mixed with inert zirconium, producing fuel elements with the desired properties.

Thorium Cycle

In the thorium fuel cycle thorium-232 absorbs a neutron in either a fast or thermal reactor. The thorium-233 beta decays to protactinium-233 and then to uranium-233, which in turn is used as fuel. Hence, like uranium-238, thorium-232 is a fertile material.

After starting the reactor with existing U-233 or some other fissile material such as U-235 or Pu-239, a breeding cycle similar to but more efficient than that with U-238 and plutonium can be created. The Th-232 absorbs a neutron to become Th-233 which quickly decays to protactinium-233. Protactinium-233 in turn decays with a half-life of 27 days to U-233. In some molten salt reactor designs, the Pa-233 is extracted and protected from neutrons (which could transform it to Pa-234 and then to U-234), until it has decayed to U-233. This is done in order to improve the breeding ratio which is low compared to fast reactors.

Thorium is at least 4-5 times more abundant in nature than all of uranium isotopes combined; thorium is fairly evenly spread around Earth with a lot of countries having huge supplies of it; preparation of thorium fuel does not require difficult and expensive enrichment processes; the thorium fuel cycle creates mainly Uranium-233 contaminated with Uranium-232 which makes it harder to use in a normal, pre-assembled nuclear weapon which is stable over long periods of time (unfortunately drawbacks are much lower for immediate use weapons or where final assembly occurs just prior to usage time); elimination of at least the transuranic portion of the nuclear waste problem is possible in MSR and other breeder reactor designs.

One of the earliest efforts to use a thorium fuel cycle took place at Oak Ridge National Laboratory in the 1960s. An experimental reactor was built based on molten salt reactor technology to study the feasibility of such an approach, using thorium fluoride salt kept hot enough to be liquid, thus eliminating the need for fabricating fuel elements. This effort culminated in the Molten-Salt Reactor Experiment that used ^{232}Th as the fertile material and ^{233}U as the fissile fuel. Due to a lack of funding, the MSR program was discontinued in 1976.

Current Industrial Activity

Currently the only isotopes used as nuclear fuel are uranium-235 (U-235), uranium-238 (U-238) and plutonium-239, although the proposed thorium fuel cycle has advantages. Some modern reactors, with minor modifications, can use thorium. Thorium is approximately three times more abundant in the Earth's crust than uranium (and 550 times more abundant than uranium-235). However, there has been little exploration for thorium resources, and thus the proved resource is small. Thorium is more plentiful than uranium in some countries, notably India.

Heavy water reactors and graphite-moderated reactors can use natural uranium, but the vast majority of the world's reactors require enriched uranium, in which the ratio of U-235 to U-238 is increased. In civilian reactors the enrichment is increased to as much as 5% U-235 and 95% U-238, but in naval reactors there is as much as 93% U-235.

The term *nuclear fuel* is not normally used in respect to fusion power, which fuses isotopes of hydrogen into helium to release energy.

Material Selection

The purpose of cladding in a nuclear reactor is of two-folds:

(i) Cladding gives the physical configuration by housing fuel pellets

(ii) Cladding retains the fission products and prevents direct contact between coolant and fuel

The material of choice for cladding must possess ductility, impact strength and creep adequate enough to maintain cladding unaltered during the operation of the reactor. The material must be suitable for fabrication in the desired form. The resistance to corrosion by coolant must be high for a material to be used as cladding. The cladding material must possess high melting point. Since cladding separates fuel and coolant, the cladding acts as additional resistance to heat transfer from fuel to coolant. To ensure rapid removal of fission heat, the material of cladding must offer low resistance for thermal conduction. In other words, the material used for cladding must possess high thermal conductivity. The material of cladding must not be damaged due to sustained neutron irradiation.

Apart from the above mechanical and thermo-physical properties, the nuclear properties of clad material are also of immense importance. This holds true especially for thermal reactors. The clad material used in thermal reactors must possess low absorption cross section for neutrons. Aluminum, beryllium, magnesium and zirconium have low absorption cross section for neutrons and possess high melting point. The mechanical properties of beryllium are poor. Moreover, beryllium is expensive. Aluminum is used in small, research reactors. Magnox, an alloy of Magnesium is used in gas-cooled reactors. Zircaloy, the alloy of zirconium is used in most thermal reactors. These alloys (Zircaloy-2 and Zircaloy-4) possess good mechanical properties and have superior resistance to corrosion.

The resistance of zirconium to corrosion is increased by addition of tin, iron, chromium and nickel. This zirconium alloy is zircaloy-2. The distribution by mass of these metals in zircaloy-2 is as follows: Tin ~ 1.2-1.7; Iron ~ 0.07-0.2; Cr ~ 0.05-0.15 and Ni ~ 0.03-0.08. By addition of these metals to zirconium, the low absorption cross section for neutrons was not comprised to a large extent. However, zircaloy-2 has tendency to react with hydrogen, an undesirable event. This problem is overcome in zircaloy-4, whose maximum nickel and iron concentrations are restricted to 0.007% and 0.12 % respectively by weight.

The requirement of low absorption cross section need not be satisfied for fast reactors. However a material with low swelling under higher fast neutron fluence is required. Stainless steel has both low swelling and good resistance to corrosion by liquid sodium. Hence stainless steel is used as cladding material in sodium cooled fast reactors.

Reflector

The purpose of reflector in a nuclear reactor is to reflect the neutrons escaping or leaving, back to the core. This serves to flatten the flux, which is essential to increase the reactor power without excessive heating of fuel elements. The following are the essential characteristics required for a reflector material:

(i) Low cross section for neutron capture or absorption

(ii) High cross section for neutron scattering

(iii) High energy loss per collision event between neutron and reflector

(iv) Temperature and radiation stability

Most thermal reactors use water as both the moderator as well as reflector. The reflector in PHWR is the heavy water. In graphite-moderated reactors, graphite acts as reflector also. These common reflector materials, which are also good moderators, cannot be used in fast reactors.

Control Elements

These elements are meant to control the reactor power through absorption of neutrons. These materials are 'poisons'. Hafnium, silver-indium-cadmium alloys and boron carbide are the widely used control materials.

The neutronic, physical and mechanical properties of Hafnium make it suitable for use as control material in water-cooled reactors. However, the availability of Hafnium is limited and hence it is expensive.

Silver-indium-cadmium alloys are as effective as Hafnium. These materials can be easily fabricated in the desired form. However, these materials must be enclosed in a stainless steel enclosure to protect the same from corrosion.

Boron has very high absorption cross section and is of low cost. However, these materials have to be incorporated in a metallic enclosure.

Shielding Materials

The types of radiations in a nuclear reactor are neutrons, gamma, alpha and beta radiation. For absorption of neutron radiation, a material with low mass number and high cross section is suitable. Water possesses the above properties, apart from being of low cost and capable of removing heat.

The material used to shield gamma radiation must be composed of high mass number element. High material density is also a desirable quality. Lead, iron and concrete are candidates for shielding gamma radiation. Lead is attractive due to its low cost. However its lower melting point is a disadvantage. Iron and concrete are good neutron absorbers. They can be easily fabricated as well. Shielding materials are not required for α and β radiation.

References

- Gudowski, W. (August 2000). "Why Accelerator-Driven Transmutation of Wastes Enables Future Nuclear Power?" (PDF). XX International Linac Conference. Retrieved 2008-01-15
- M. I. Ojovan, W.E. Lee. An Introduction to Nuclear Waste Immobilisation, Elsevier Science Publishers B.V., ISBN 0-08-044462-8, Amsterdam, 315pp. (2005)
- Milsted, J.; Friedman, A. M.; Stevens, C. M. (1965). "The alpha half-life of berkelium-247; a new long-lived isomer of berkelium-248". Nuclear Physics. 71 (2): 299. doi:10.1016/0029-5582(65)90719-4
- Brolly Á.; Vértes P. (March 2005). "Concept of a Small-scale Electron Accelerator Driven System for Nuclear Waste Transmutation Part 2. Investigation of burnup" (PDF). Retrieved 2008-01-15
- Harvey, L.D.D. (2010). Energy and the New Reality 2: Carbon-Free Energy Supply- section 8.4. Earthscan. ISBN 9781849710732
- "Accelerator-driven Systems (ADS) and Fast Reactors (FR) in Advanced Nuclear Fuel Cycles" (PDF). Nuclear Energy Agency. Retrieved 2008-01-15

3

Physical Components of Nuclear Reactors

The design constraints related to nuclear reactors are the maximum temperature of clad, linear (heat/power) rating, coolant velocity, etc. Corrosion is also an important factor to be kept in mind while designing a nuclear reactor. This chapter elucidates the crucial theories and principles of nuclear science and technology.

Reactor Design Limitations

Some of the design constraints are listed below with their significance and their influence on reactor characteristics:

Maximum Temperature of Clad

From the thermodynamic point of view, higher temperature of clad is advantageous. This would in turn increase the outlet temperature of primary sodium, and in turn that of secondary sodium and in turn that of steam. The efficiency of thermal to electrical energy conversion increases with increase in steam temperature.

However the properties of clad material impose constraints on the maximum permissible clad temperature. Another problem in achieving higher clad temperature is the resistance for heat conduction within the pellet and that in the fuel gap.

As discussed earlier the thermal conductivity of mixed oxide fuel is low. Hence to achieve a fixed rate of heat transfer, the temperature gradient required is high. In other terms the difference in temperature between the centre of fuel pin and that at the outer surface will be high. If one considers the resistance in the fuel-clad gap and in the clad

also, the difference between the temperature at the centre of the fuel pin and the temperature at the outer surface of the clad will be very high.

The net result of all these resistances is that to obtain higher temperature on the outer surface of the clad, higher temperature at the centre of fuel pin is required. If this temperature exceeds the melting point of the fuel, catastrophic consequences would set in. Hence, this too limits the maximum temperature on the outer surface of the cladding.

Under steady-state condition, the cladding material is exposed to temperatures in the range of 673-973 K. However during transient conditions, the temperature may reach 1273 K. One can understand this, by looking at the following examples.

Example – 1: The average temperature of coolant in fast breeder reactor is 470°C. If the coolant-side heat transfer coefficient is 15000 W/m²K, determine the temperature on the outer surface of the cladding, if the diameter of pin is 6 mm and cladding is 0.5 mm thick. The rate of heat generation is 1000 MW/m³.

Data: Volumetric rate of heat generation = q' = $1x10^9$ W/m³

Converting 'q' in terms of per unit surface area of pins, we get heat flux (q") as follows:

q"(surface area of pins) = q' (Volume of pins)

$q''(NL2\pi R)=q'(NL\pi R^2)$

$q'' = q'R/2$

Heat flux (q") may also be equated to the heat transfer coefficient (h) and driving force (T_{clad}-$T_{coolant}$) as follows:

$q''=q'R/2=h(T_{clad}-T_{coolant})$

Substituting R = 3+0.5 = 3.5 mm = $3.5x10^{-3}$ m, h = 15000 W/m²K and q'= 1×10^9W/m³, we get

$T_{clad} = T_{coolant}+q'R/(2h)$

$T_{clad} = T_{coolant}+q''/h = 587$ °C

Therefore, the temperature on the outer surface of cladding is 587 °C.

Example – 2: In the above problem if the heat transfer coefficient falls to 5000 W/m²K due to reduced coolant flow, determine the temperature on the outer surface of the cladding.

Solution:

In the above problem, if the value of 'h' is substituted as 5000 W/m²K, we get

$T_{clad} = 820$ °C

This problem illustrates the increased temperature of clad due to transients in the coolant flow rate.

Linear (Heat/Power) Rating

It is defined as the ratio of thermal power to the product of number of fuel pins and the height (or length) of the pin.

$$Linear\ rating = \frac{Q}{N*L} \qquad (1)$$

In terms of thermal conductivity of the material, linear rating is expressed as

$$Linear\ rating = 4\pi\int_{Ts}^{Tfc} k(T).dT \qquad (2)$$

In Eq. (2), T_s is the temperature on fuel surface and T_{fc} is the temperature at the fuel centre. k(T) is the temperature-dependent thermal conductivity of the fuel.

From the above expression, one may observe that the linear power rating depends on the thermal conductivity of the material and its variation with temperature. Higher linear rating will be obtained for materials that have higher thermal conductivity. This is illustrated in the following examples:

Determine the linear rating of an oxide fuel when the temperature at the center of the fuel pin is 800 °C while that at the surface is 600 °C. The thermal conductivity of oxide fuel is related to temperature (in K) as follows:

$$k(T) = \frac{3824}{129.4+T} + 6.125x10^{-11}\left(T^3\right); \frac{W}{mK}$$

$$Linear\ rating = 4\pi \int_{600+273}^{800+273} \frac{3824}{129.4+T} + 6.125x10^{-11}\left(T^3\right)dT$$

$$Linear\ rating = 4\pi\left[3824\ln\left(129.4+T\right) + 6.125x10^{-11}\left(\frac{T^4}{4}\right)\right]_{873}^{1073}$$

From the above equation, Linear rating =8885.4 W/m

Assuming that a metallic fuel has a linear rating of 8885.4 W/m (as calculated above), determine the temperature at the centre of the fuel if the temperature on its outer surface is 600°C. The thermal conductivity of metallic fuel as the function of temperature (in K) is given by

$$k(T) = 6.945681 + 0.016174(T) + 5.58x10^{-6}(T^2); W/mK$$

Solution:

$$Linear\, rating = 4\pi \int_{600+273}^{T} 6.945681 + 0.016174(T) + 5.58x10^{-6}(T^2)dT$$

$$Linear\, rating = 4\pi\left[6.945681(T) + 0.016174T^2/2 + \frac{5.58x10^{-6}(T^3)}{3}\right]_{873}^{T}$$

Solving the above equation, we get T = 900 K

What do we infer from Examples – 1 and 2? For the same linear rating, with metallic fuels the temperature at the centre of the fuel is lower than that of a metal oxide fuel. In other words, the linear power of a metallic fuel can be increased appreciably before the peak fuel temperature or temperature at the centre of the fuel increases beyond permissible limits.

Since linear rating and reactor thermal power are directly proportional, a fuel element with higher linear rating can be used to produce higher thermal power for the same number of fuel pins and for the same length.

Once it is realized that the higher linear rating is advantageous, it must also be realized that for a chosen fuel material, larger difference in temperature between the fuel center and fuel surface would be incurred for higher linear rating. As discussed earlier, there is a limitation to maximum temperature to which a fuel can be heated. This poses a constraint on the maximum linear rating permissible, depending on the type of fuel used. Use of large number of small diameter fuel pins can be used to increase the reactor power without increasing the linear rating. The maximum linear rating for PFBR is 450 W/cm with an active core height of 1 m.

Heat Transfer Coefficient

The heat transfer coefficient or film coefficient, or film effectiveness, in thermodynamics and in mechanics is the proportionality constant between the heat flux and the thermodynamic driving force for the flow of heat (i.e., the temperature difference, ΔT):

$$h = \frac{q}{\Delta T}$$

where:

q: amount of heat transferred (heat flux), W/m^2 i.e., thermal power per unit area, $q = d\dot{Q}/dA$

h: heat transfer coefficient, W/(m^2•K)

ΔT: difference in temperature between the solid surface and surrounding fluid area, K.

It is used in calculating the heat transfer, typically by convection or phase transition between a fluid and a solid. The heat transfer coefficient has SI units in watts per squared meter kelvin: W/(m^{2}K).

The heat transfer coefficient is the reciprocal of thermal insulance. This is used for building materials (R-value) and for clothing insulation.

There are numerous methods for calculating the heat transfer coefficient in different heat transfer modes, different fluids, flow regimes, and under different thermohydraulic conditions. Often it can be estimated by dividing the thermal conductivity of the convection fluid by a length scale. The heat transfer coefficient is often calculated from the Nusselt number (a dimensionless number). There are also online calculators available specifically for heat transfer fluid applications. Experimental assessment of the heat transfer coefficient poses some challenges especially when small fluxes are to be measured (e.g. $< 0.2W/cm^2$).

Composition

A simple method for determining an overall heat transfer coefficient that is useful to find the heat transfer between simple elements such as walls in buildings or across heat exchangers is shown below. Note that this method only accounts for conduction within materials, it does not take into account heat transfer through methods such as radiation. The method is as follows:

$$1/(U \cdot A) = 1/(h_1 \cdot A_1) + dx_w/(k \cdot A) + 1/(h_2 \cdot A_2)$$

Where:

- U = the overall heat transfer coefficient (W/(m^2•K))

- A = the contact area for each fluid side (m²) (with A_1 and A_2 expressing either surface)
- k = the thermal conductivity of the material (W/(m·K))
- h = the individual convection heat transfer coefficient for each fluid (W/(m²•K))
- dx_w = the wall thickness (m).

As the areas for each surface approach being equal the equation can be written as the transfer coefficient per unit area as shown below:

$$1/U = 1/h_1 + dx_w/k + 1/h_2$$

or

$$U = 1/\left(1/h_1 + dx_w/k + 1/h_2\right)$$

It is to be noted that often the value for dx_w is referred to as the difference of two radii where the inner and outer radii are used to define the thickness of a pipe carrying a fluid, however, this figure may also be considered as a wall thickness in a flat plate transfer mechanism or other common flat surfaces such as a wall in a building when the area difference between each edge of the transmission surface approaches zero.

In the walls of buildings the above formula can be used to derive the formula commonly used to calculate the heat through building components. Architects and engineers call the resulting values either the U-Value or the R-Value of a construction assembly like a wall. Each type of value (R or U) are related as the inverse of each other such that R-Value = 1/U-Value and both are more fully understood through the concept of an overall heat transfer coefficient described in lower section of this document.

Convective Heat Transfer Correlations

Although convective heat transfer can be derived analytically through dimensional analysis, exact analysis of the boundary layer, approximate integral analysis of the boundary layer and analogies between energy and momentum transfer, these analytic approaches may not offer practical solutions to all problems when there are no mathematical models applicable. Therefore, many correlations were developed by various authors to estimate the convective heat transfer coefficient in various cases including natural convection, forced convection for internal flow and forced convection for external flow. These empirical correlations are presented for their particular geometry and flow conditions. As the fluid properties are temperature dependent, they are evaluated at the film temperature T_f, which is the average of the surface T_s and the surrounding bulk temperature, T_∞.

$$T_f = \frac{T_s + T_\infty}{2}$$

External Flow, Vertical Plane

Recommendations by Churchill and Chu provide the following correlation for natural convection adjacent to a vertical plane, both for laminar and turbulent flow. k is the thermal conductivity of the fluid, L is the characteristic length with respect to the direction of gravity, Ra_L is the Rayleigh number with respect to this length and Pr is the Prandtl number.

$$h = \frac{k}{L}\left(0.825 + \frac{0.387 Ra_L^{1/6}}{\left(1 + (0.492 / Pr)^{9/16}\right)^{8/27}}\right)^2 \qquad Ra_L < 10^{12}$$

For laminar flows, the following correlation is slightly more accurate. It is observed that a transition from a laminar to a turbulent boundary occurs when Ra_L exceeds around 10^9.

$$h = \frac{k}{L}\left(0.68 + \frac{0.67 Ra_L^{1/4}}{\left(1 + (0.492 / Pr)^{9/16}\right)^{4/9}}\right) \qquad 10^{-1} < Ra_L < 10^9$$

External Flow, Vertical Cylinders

For cylinders with their axes vertical, the expressions for plane surfaces can be used provided the curvature effect is not too significant. This represents the limit where boundary layer thickness is small relative to cylinder diameter D. The correlations for vertical plane walls can be used when

$$\frac{D}{L} \geq \frac{35}{Gr_L^{\frac{1}{4}}}$$

where Gr_L is the Grashof number.

External Flow, Horizontal Plates

W. H. McAdams suggested the following correlations for horizontal plates. The induced buoyancy will be different depending upon whether the hot surface is facing up or down.

For a hot surface facing up, or a cold surface facing down, for laminar flow:

$$h = \frac{k 0.54 Ra_L^{1/4}}{L} \qquad 10^5 < Ra_L < 2\times10^7$$

and for turbulent flow:

$$h = \frac{k 0.14 Ra_L^{1/3}}{L} \qquad 2\times10^7 < Ra_L < 3\times10^{10}.$$

For a hot surface facing down, or a cold surface facing up, for laminar flow:

$$h = \frac{k 0.27 Ra_L^{1/4}}{L} \qquad 3\times10^5 < Ra_L < 3\times10^{10}.$$

The characteristic length is the ratio of the plate surface area to perimeter. If the surface is inclined at an angle θ with the vertical then the equations for a vertical plate by Churchill and Chu may be used for θ up to 60°; if the boundary layer flow is laminar, the gravitational constant g is replaced with $g\cos\theta$ when calculating the Ra term.

External Flow, Horizontal Cylinder

For cylinders of sufficient length and negligible end effects, Churchill and Chu has the following correlation for $10^{-5} < \mathrm{Ra}_D < 10^{12}$.

$$h = \frac{k}{D}\left(0.6 + \frac{0.387 Ra_D^{1/6}}{\left(1+(0.559/Pr)^{9/16}\right)^{8/27}}\right)^2$$

External Flow, Spheres

For spheres, T. Yuge has the following correlation for Pr≃1 and $1 \le Ra_D \le 10^5$.

$$Nu_D = 2 + 0.43 Ra_D^{1/4}$$

Forced Convection

Internal Flow, Laminar Flow

Sieder and Tate has the following correlation for laminar flow in tubes where D is the internal diameter, μ_b is the fluid viscosity at the bulk mean temperature, μ_w is the viscosity at the tube wall surface temperature.

$$Nu_D = 1.86 \cdot (Re \cdot Pr)^{1/3} \left(\frac{D}{L}\right)^{1/3} \left(\frac{\mu_b}{\mu_w}\right)^{0.14}$$

Internal Flow, Turbulent Flow

The Dittus-Bölter correlation (1930) is a common and particularly simple correlation useful for many applications. This correlation is applicable when forced convection is the only mode of heat transfer; i.e., there is no boiling, condensation, significant radiation, etc. The accuracy of this correlation is anticipated to be ±15%.

For a fluid flowing in a straight circular pipe with a Reynolds number between 10,000 and 120,000 (in the turbulent pipe flow range), when the fluid's Prandtl number is between 0.7 and 120, for a location far from the pipe entrance (more than 10 pipe diameters; more than 50 diameters according to many authors) or other flow disturbances, and when the pipe surface is hydraulically smooth, the heat transfer coefficient between the bulk of the fluid and the pipe surface can be expressed explicitly as:

$$\frac{hd}{k} = 0.023\left(\frac{jd}{\mu}\right)^{0.8}\left(\frac{\mu c_p}{k}\right)^n$$

where:

d is the hydraulic diameter

k is the thermal conductivity of the bulk fluid

μ viscosity

j mass flux

isobaric heat capacity

n = 0.4 for heating (wall hotter than the bulk fluid) and 0.33 for cooling (wall cooler than the bulk fluid).

The fluid properties necessary for the application of this equation are evaluated at the bulk temperature thus avoiding iteration.

Forced Convection, External Flow

In analyzing the heat transfer associated with the flow past the exterior surface of a solid, the situation is complicated by phenomena such as boundary layer separation. Various authors have correlated charts and graphs for different geometries and flow conditions. For flow parallel to a plane surface, where x is the distance from the edge and L is the height of the boundary layer, a mean Nusselt number can be calculated using the Colburn analogy.

Thom Correlation

There exist simple fluid-specific correlations for heat transfer coefficient in boiling. The Thom correlation is for the flow of boiling water (subcooled or saturated at pressures up to about 20 MPa) under conditions where the nucleate boiling contribution predominates over forced convection. This correlation is useful for rough estimation of expected temperature difference given the heat flux:

$$\Delta T_{sat} = 22.5 \cdot q^{0.5} \exp(-P/8.7)$$

where:

ΔT_{sat} is the wall temperature elevation above the saturation temperature, K

q is the heat flux, MW/m²

P is the pressure of water, MPa

Note that this empirical correlation is specific to the units given.

Heat Transfer Coefficient of Pipe Wall

The resistance to the flow of heat by the material of pipe wall can be expressed as a "heat transfer coefficient of the pipe wall". However, one needs to select if the heat flux is based on the pipe inner or the outer diameter. Selecting to base the heat flux on the pipe inner diameter, and assuming that the pipe wall thickness is small in comparison with the pipe inner diameter, then the heat transfer coefficient for the pipe wall can be calculated as if the wall were not curved:

$$h_{wall} = \frac{k}{x}$$

where k is the effective thermal conductivity of the wall material and x is the wall thickness.

If the above assumption does not hold, then the wall heat transfer coefficient can be calculated using the following expression:

$$h_{wall} = \frac{2k}{d_i \ln(d_o / d_i)}$$

where d_i and d_o are the inner and outer diameters of the pipe, respectively.

The thermal conductivity of the tube material usually depends on temperature; the mean thermal conductivity is often used.

Combining Heat Transfer Coefficients

For two or more heat transfer processes acting in parallel, heat transfer coefficients simply add:

$$h = h_1 + h_2 + \cdots$$

For two or more heat transfer processes connected in series, heat transfer coefficients add inversely:

$$\frac{1}{h} = \frac{1}{h_1} + \frac{1}{h_2} + \ldots$$

For example, consider a pipe with a fluid flowing inside. The approximate rate of heat transfer between the bulk of the fluid inside the pipe and the pipe external surface is:

$$q = \left(\frac{1}{\frac{1}{h} + \frac{t}{k}} \right) \cdot A \cdot \Delta T$$

where

q = heat transfer rate (W)

h = heat transfer coefficient (W/(m^2·K))

t = wall thickness (m)

k = wall thermal conductivity (W/m·K)

A = area (m^2)

ΔT = difference in temperature.

Overall Heat Transfer Coefficient

The overall heat transfer coefficient U is a measure of the overall ability of a series of conductive and convective barriers to transfer heat. It is commonly applied to the calculation of heat transfer in heat exchangers, but can be applied equally well to other problems.

For the case of a heat exchanger, U can be used to determine the total heat transfer between the two streams in the heat exchanger by the following relationship:

$$q = UA\Delta T_{LM}$$

where:

q = heat transfer rate (W)

U = overall heat transfer coefficient (W/(m^2·K))

A = heat transfer surface area (m^2)

ΔT_{LM} = logarithmic mean temperature difference (K).

The overall heat transfer coefficient takes into account the individual heat transfer coefficients of each stream and the resistance of the pipe material. It can be calculated as the reciprocal of the sum of a series of thermal resistances (but more complex relationships exist, for example when heat transfer takes place by different routes in parallel):

$$\frac{1}{UA} = \sum \frac{1}{hA} + \sum R$$

where:

R = Resistance(s) to heat flow in pipe wall (K/W)

Other parameters are as above.

The heat transfer coefficient is the heat transferred per unit area per kelvin. Thus *area* is included in the equation as it represents the area over which the transfer of heat takes place. The areas for each flow will be different as they represent the contact area for each fluid side.

The *thermal resistance* due to the pipe wall is calculated by the following relationship:

$$R = \frac{x}{k \cdot A}$$

where

x = the wall thickness (m)

k = the thermal conductivity of the material (W/(m·K))

A = the total area of the heat exchanger (m^2)

This represents the heat transfer by conduction in the pipe.

The *thermal conductivity* is a characteristic of the particular material. Values of thermal conductivities for various materials are listed in the list of thermal conductivities.

The *convection heat transfer coefficient* for each stream depends on the type of fluid, flow properties and temperature properties.

Some typical heat transfer coefficients include:

- Air - h = 10 to 100 W/(m²K)
- Water - h = 500 to 10,000 W/(m²K).

Thermal Resistance due to Fouling Deposits

Often during their use, heat exchangers collect a layer of fouling on the surface which, in addition to potentially contaminating a stream, reduces the effectiveness of heat exchangers. In a fouled heat exchanger the buildup on the walls creates an additional layer of materials that heat must flow through. Due to this new layer, there is additional resistance within the heat exchanger and thus the overall heat transfer coefficient of the exchanger is reduced. The following relationship is used to solve for the heat transfer resistance with the additional fouling resistance:

$$\frac{1}{U_f P} = \frac{1}{UP} + \frac{R_{fH}}{P_H} + \frac{R_{fC}}{P_C}$$

where

U_f = overall heat transfer coefficient for a fouled heat exchanger, $\frac{W}{m^2K}$

P = perimeter of the heat exchanger, may be either the hot or cold side perimeter however, it must be the same perimeter on both sides of the equation, m

U = overall heat transfer coefficient for an unfouled heat exchanger, $\frac{W}{m^2K}$

R_{fC} = fouling resistance on the cold side of the heat exchanger, $\frac{m^2K}{W}$

R_{fH} = fouling resistance on the hot side of the heat exchanger, $\frac{m^2K}{W}$

P_C = perimeter of the cold side of the heat exchanger, m

P_H = perimeter of the hot side of the heat exchanger, m

This equation uses the overall heat transfer coefficient of an unfouled heat exchanger and the fouling resistance to calculate the overall heat transfer coefficient of a fouled heat exchanger. The equation takes into account that the perimeter of the heat exchanger is different on the hot and cold sides. The perimeter used for the P does not matter as long as it is the same. The overall heat transfer coefficients will adjust to take into account that a different perimeter was used as the product UP will remain the same.

The fouling resistances can be calculated for a specific heat exchanger if the average thickness and thermal conductivity of the fouling are known. The product of the average thickness and thermal conductivity will result in the fouling resistance on a specific side of the heat exchanger.

$$R_f = \frac{d_f}{k_f}$$

where:

d_f = average thickness of the fouling in a heat exchanger, m

k_f = thermal conductivity of the fouling, $\frac{W}{mK}$.

Corrosion

Rust, the most familiar example of corrosion

Volcanic gases have accelerated the extensive corrosion of this abandoned mining machinery, rendering it almost unrecognizable

Corrosion on exposed metal, including a bolt and nut

Corrosion is a natural process, which converts a refined metal to a more chemically-stable form, such as its oxide, hydroxide, or sulfide. It is the gradual destruction of materials (usually metals) by chemical and/or electrochemical reaction with their environment. Corrosion engineering is the field dedicated to controlling and stopping corrosion.

In the most common use of the word, this means electrochemical oxidation of metal in reaction with an oxidant such as oxygen or sulfur. Rusting, the formation of iron oxides, is a well-known example of electrochemical corrosion. This type of damage typically produces oxide(s) or salt(s) of the original metal, and results in a distinctive orange colouration. Corrosion can also occur in materials other than metals, such as ceramics or polymers, although in this context, the term "degradation" is more common. Corrosion degrades the useful properties of materials and structures including strength, appearance and permeability to liquids and gases.

Many structural alloys corrode merely from exposure to moisture in air, but the process can be strongly affected by exposure to certain substances. Corrosion can be concentrated locally to form a pit or crack, or it can extend across a wide area more or less uniformly corroding the surface. Because corrosion is a diffusion-controlled process, it occurs on exposed surfaces. As a result, methods to reduce the activity of the exposed surface, such as passivation and chromate conversion, can increase a material's corrosion resistance. However, some corrosion mechanisms are less visible and less predictable.

Galvanic Corrosion

Galvanic corrosion of aluminium. A 5-mm-thick aluminium alloy plate is physically (and hence, electrically) connected to a 10-mm-thick mild steel structural support. Galvanic corrosion occurred on the aluminium plate along the joint with the steel. Perforation of aluminium plate occurred within 2 years.

Galvanic corrosion occurs when two different metals have physical or electrical contact with each other and are immersed in a common electrolyte, or when the same metal is exposed to electrolyte with different concentrations. In a galvanic couple, the more active metal (the anode) corrodes at an accelerated rate and the more noble metal (the cathode) corrodes at a slower rate. When immersed separately, each metal corrodes at its own rate. What type of metal(s) to use is readily determined by following the galvanic series. For example, zinc is often used as a sacrificial anode for steel structures. Galvanic corrosion is of major interest to the marine industry and also anywhere water (containing salts) contacts pipes or metal structures.

Factors such as relative size of anode, types of metal, and operating conditions (temperature, humidity, salinity, etc.) affect galvanic corrosion. The surface area ratio of the anode and cathode directly affects the corrosion rates of the materials. Galvanic corrosion is often prevented by the use of sacrificial anodes.

Galvanic Series

In any given environment (one standard medium is aerated, room-temperature seawater), one metal will be either more *noble* or more *active* than others, based on how strongly its ions are bound to the surface. Two metals in electrical contact share the same electrons, so that the "tug-of-war" at each surface is analogous to competition for free electrons between the two materials. Using the electrolyte as a host for the flow of ions in the same direction, the noble metal will take electrons from the active one. The resulting mass flow or electric current can be measured to establish a hierarchy of materials in the medium of interest. This hierarchy is called a *galvanic series* and is useful in predicting and understanding corrosion.

Corrosion Removal

Often it is possible to chemically remove the products of corrosion. For example, phosphoric acid in the form of naval jelly is often applied to ferrous tools or surfaces to remove rust. Corrosion removal should not be confused with electropolishing, which removes some layers of the underlying metal to make a smooth surface. For example, phosphoric acid may also be used to electropolish copper but it does this by removing copper, not the products of copper corrosion.

Resistance to Corrosion

Some metals are more intrinsically resistant to corrosion than others. There are various ways of protecting metals from corrosion (oxidation) including painting, hot dip galvanizing, and combinations of these.

Intrinsic Chemistry

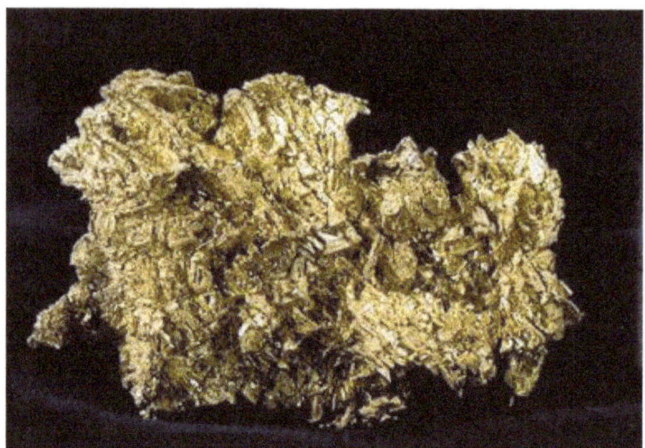

Gold nuggets do not naturally corrode, even on a geological time scale.

The materials most resistant to corrosion are those for which corrosion is thermodynamically unfavorable. Any corrosion products of gold or platinum tend to decompose spontaneously into pure metal, which is why these elements can be found in metallic form on Earth and have long been valued. More common "base" metals can only be protected by more temporary means.

Some metals have naturally slow reaction kinetics, even though their corrosion is thermodynamically favorable. These include such metals as zinc, magnesium, and cadmium. While corrosion of these metals is continuous and ongoing, it happens at an acceptably slow rate. An extreme example is graphite, which releases large amounts of energy upon oxidation, but has such slow kinetics that it is effectively immune to electrochemical corrosion under normal conditions.

Passivation

Passivation refers to the spontaneous formation of an ultrathin film of corrosion products, known as a passive film, on the metal's surface that act as a barrier to further oxidation. The chemical composition and microstructure of a passive film are different from the underlying metal. Typical passive film thickness on aluminium, stainless steels, and alloys is within 10 nanometers. The passive film is different from oxide layers that are formed upon heating and are in the micrometer thickness range – the passive film recovers if removed or damaged whereas the oxide layer does not. Passivation in natural environments such as air, water and soil at moderate pH is seen in such materials as aluminium, stainless steel, titanium, and silicon.

Passivation is primarily determined by metallurgical and environmental factors. The effect of pH is summarized using Pourbaix diagrams, but many other factors are influential. Some conditions that inhibit passivation include high pH for aluminium and zinc, low pH or the presence of chloride ions for stainless steel, high temperature for titanium (in which case the oxide dissolves into the metal, rather than the electrolyte) and fluoride ions for silicon. On the other hand, unusual conditions may result in passivation of materials that are normally unprotected, as the alkaline environment of concrete does for steel rebar. Exposure to a liquid metal such as mercury or hot solder can often circumvent passivation mechanisms.

Corrosion in Passivated Materials

Passivation is extremely useful in mitigating corrosion damage, however even a high-quality alloy will corrode if its ability to form a passivating film is hindered. Proper selection of the right grade of material for the specific environment is important for the long-lasting performance of this group of materials. If breakdown occurs in the passive film due to chemical or mechanical factors, the resulting major modes of corrosion may include pitting corrosion, crevice corrosion, and stress corrosion cracking.

Pitting Corrosion

Certain conditions, such as low concentrations of oxygen or high concentrations of species such as chloride which complete as anions, can interfere with a given alloy's ability to re-form a passivating film. In the worst case, almost all of the surface will remain protected, but tiny local fluctuations will degrade the oxide film in a few critical points. Corrosion at these points will be greatly amplified, and can cause *corrosion pits* of several types, depending upon conditions. While the corrosion pits only nucleate under fairly extreme circumstances, they can continue to grow even when conditions return to normal, since the interior of a pit is naturally deprived of oxygen and locally the pH decreases to very low values and the corrosion rate increases due to an autocatalytic process. In extreme cases, the sharp tips of extremely long and narrow corrosion pits can cause stress concentration to the point that otherwise tough alloys can shatter; a thin film pierced by an invisibly small hole can hide a thumb sized pit from view. These problems are especially dangerous because they are difficult to detect before a part or structure fails. Pitting remains among the most common and damaging forms of corrosion in passivated alloys, but it can be prevented by control of the alloy's environment.

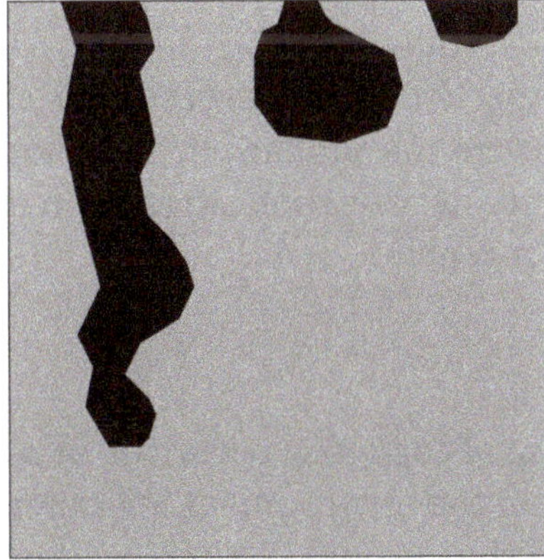

Diagram showing cross-section of pitting corrosion

Pitting results when a small hole, or cavity, forms in the metal, usually as a result of de-passivation of a small area. This area becomes anodic, while part of the remaining metal becomes cathodic, producing a localized galvanic reaction. The deterioration of this small area penetrates the metal and can lead to failure. This form of corrosion is often difficult to detect due to the fact that it is usually relatively small and may be covered and hidden by corrosion-produced compounds.

Weld Decay and Knifeline Attack

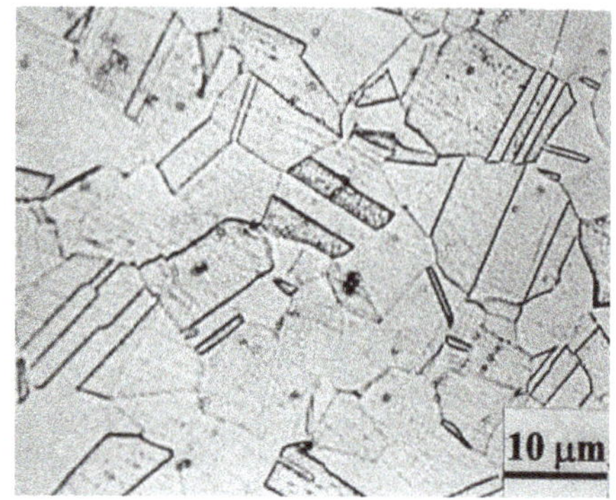

Normal microstructure of Type 304 stainless steel surface

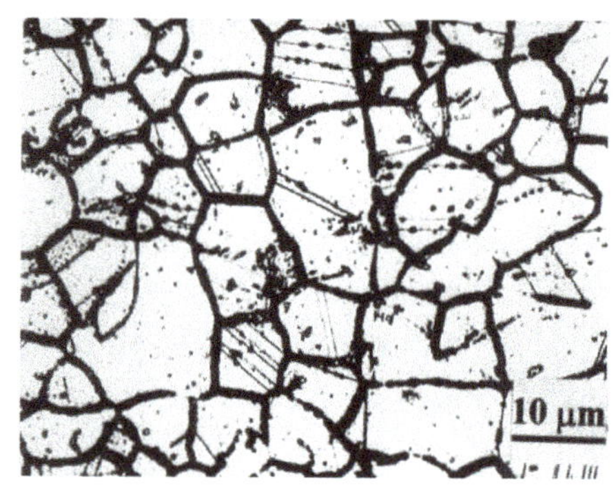

Sensitized metallic microstructure, showing wider intergranular boundaries

Stainless steel can pose special corrosion challenges, since its passivating behavior relies on the presence of a major alloying component (chromium, at least 11.5%). Because of the elevated temperatures of welding and heat treatment, chromium carbides can form in the grain boundaries of stainless alloys. This chemical reaction robs the material of chromium in the zone near the grain boundary, making those areas much less resistant to corrosion. This creates a galvanic couple with the well-protected alloy nearby, which leads to "weld decay" (corrosion of the grain boundaries in the heat affected zones) in highly corrosive environments. This process can seriously reduce the mechanical strength of welded joints over time.

A stainless steel is said to be "sensitized" if chromium carbides are formed in the microstructure. A typical microstructure of a normalized type 304 stainless steel shows no signs of sensitization, while a heavily sensitized steel shows the presence of grain boundary precipitates. The dark lines in the sensitized microstructure are networks of chromium carbides formed along the grain boundaries.

Special alloys, either with low carbon content or with added carbon "getters" such as titanium and niobium (in types 321 and 347, respectively), can prevent this effect, but the latter require special heat treatment after welding to prevent the similar phenomenon of "knifeline attack". As its name implies, corrosion is limited to a very narrow zone adjacent to the weld, often only a few micrometers across, making it even less noticeable.

Crevice Corrosion

Crevice corrosion is a localized form of corrosion occurring in confined spaces (crevices), to which the access of the working fluid from the environment is limited. Formation of a differential aeration cell leads to corrosion inside the crevices. Examples of crevices are gaps and contact areas between parts, under gaskets or seals, inside cracks and seams, spaces filled with deposits and under sludge piles.

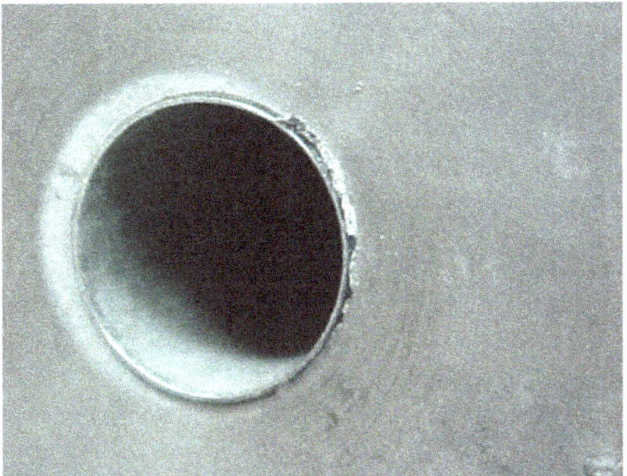

Corrosion in the crevice between the tube and tube sheet (both made of type 316 stainless steel) of a heat exchanger in a seawater desalination plant

Crevice corrosion is influenced by the crevice type (metal-metal, metal-nonmetal), crevice geometry (size, surface finish), and metallurgical and environmental factors. The susceptibility to crevice corrosion can be evaluated with ASTM standard procedures. A critical crevice corrosion temperature is commonly used to rank a material's resistance to crevice corrosion.

Microbial Corrosion

Microbial corrosion, or commonly known as microbiologically influenced corrosion (MIC), is a corrosion caused or promoted by microorganisms, usually chemoautotrophs. It can apply to both metallic and non-metallic materials, in the presence or absence of oxygen. Sulfate-reducing bacteria are active in the absence of oxygen (anaerobic); they produce hydrogen sulfide, causing sulfide stress cracking. In the presence of oxygen (aerobic), some bacteria may directly oxidize iron to iron oxides and hydroxides, other bacteria oxidize sulfur and produce sulfuric acid causing biogenic sulfide corrosion. Concentration cells can form in the deposits of corrosion products, leading to localized corrosion.

Accelerated low-water corrosion (ALWC) is a particularly aggressive form of MIC that affects steel piles in seawater near the low water tide mark. It is characterized by an orange sludge, which smells of hydrogen sulfide when treated with acid. Corrosion rates can be very high and design corrosion allowances can soon be exceeded leading to premature failure of the steel pile. Piles that have been coated and have cathodic protection installed at the time of construction are not susceptible to ALWC. For unprotected piles, sacrificial anodes can be installed local to the affected areas to inhibit the corrosion or a complete retrofitted sacrificial anode system can be installed. Affected areas can also be treated using cathodic protection, using either sacrificial anodes or applying current to an inert anode to produce a calcareous deposit, which will help shield the metal from further attack.

High-temperature Corrosion

High-temperature corrosion is chemical deterioration of a material (typically a metal) as a result of heating. This non-galvanic form of corrosion can occur when a metal is subjected to a hot atmosphere containing oxygen, sulfur, or other compounds capable of oxidizing (or assisting the oxidation of) the material concerned. For example, materials used in aerospace, power generation and even in car engines have to resist sustained periods at high temperature in which they may be exposed to an atmosphere containing potentially highly corrosive products of combustion.

The products of high-temperature corrosion can potentially be turned to the advantage of the engineer. The formation of oxides on stainless steels, for example, can provide a protective layer preventing further atmospheric attack, allowing for a material to be used for sustained periods at both room and high temperatures in hostile conditions. Such high-temperature corrosion products, in the form of compacted oxide layer glazes, prevent or reduce wear during high-temperature sliding contact of metallic (or metallic and ceramic) surfaces.

Metal Dusting

Metal dusting is a catastrophic form of corrosion that occurs when susceptible materials are exposed to environments with high carbon activities, such as synthesis gas and other high-CO environments. The corrosion manifests itself as a break-up of bulk metal to metal powder. The suspected mechanism is firstly the deposition of a graphite layer on the surface of the metal, usually from carbon monoxide (CO) in the vapor phase. This graphite layer is then thought to form metastable M_3C species (where M is the metal), which migrate away from the metal surface. However, in some regimes no M_3C species is observed indicating a direct transfer of metal atoms into the graphite layer.

Protection from Corrosion

Various treatments are used to slow corrosion damage to metallic objects which are exposed to the weather, salt water, acids, or other hostile environments. Some unprotected metallic alloys are extremely vulnerable to corrosion, such as those used in neodymium magnets, which can spall or crumble into powder even in dry, temperature-stable indoor environments unless properly treated to discourage corrosion.

The US Army shrink wraps equipment such as helicopters to protect them from corrosion and thus save millions of dollars

Surface Treatments

When surface treatments are used to retard corrosion, great care must be taken to ensure complete coverage, without gaps, cracks, or pinhole defects. Small defects can act as an “Achilles’ heel”, allowing corrosion to penetrate the interior and causing extensive damage even while the outer protective layer remains apparently intact for a period of time.

Applied Coatings

Galvanized surface

Plating, painting, and the application of enamel are the most common anti-corrosion treatments. They work by providing a barrier of corrosion-resistant material between the damaging environment and the structural material. Aside from cosmetic and manufacturing issues, there may be tradeoffs in mechanical flexibility versus resistance to abrasion and high temperature. Platings usually fail only in small sections, but if the plating is more noble than the substrate (for example, chromium on steel), a galvanic couple will cause any exposed area to corrode much more rapidly than an unplated surface would. For this reason, it is often wise to plate with active metal such as zinc or cadmium.

Painting either by roller or brush is more desirable for tight spaces; spray would be better for larger coating areas such as steel decks and waterfront applications. Flexible polyurethane coatings, like Durabak-M26 for example, can provide an anti-corrosive seal with a highly durable slip resistant membrane. Painted coatings are relatively easy to apply and have fast drying times although temperature and humidity may cause dry times to vary.

Reactive Coatings

If the environment is controlled (especially in recirculating systems), corrosion inhibitors can often be added to it. These chemicals form an electrically insulating or chemically impermeable coating on exposed metal surfaces, to suppress electrochemical reactions. Such methods make the system less sensitive to scratches or defects in the coating, since extra inhibitors can be made available wherever metal becomes exposed. Chemicals that inhibit corrosion include some of the salts in hard water (Roman water systems are famous for their mineral deposits), chromates, phosphates, polyaniline, other conducting polymers and a wide range of specially-designed chemicals that resemble surfactants (i.e. long-chain organic molecules with ionic end groups).

Anodization

Aluminium alloys often undergo a surface treatment. Electrochemical conditions in the bath are carefully adjusted so that uniform pores, several nanometers wide, appear in the metal's oxide film. These pores allow the oxide to grow much thicker than passivating conditions would allow. At the end of the treatment, the pores are allowed to seal, forming a harder-than-usual surface layer. If this coating is scratched, normal passivation processes take over to protect the damaged area.

This climbing descender is anodized with a yellow finish.

Anodizing is very resilient to weathering and corrosion, so it is commonly used for building facades and other areas where the surface will come into regular contact with the elements. While being resilient, it must be cleaned frequently. If left without cleaning, panel edge staining will naturally occur. Anodization is the process of converting an anode into cathode by bringing a more active anode in contact with it.

Biofilm Coatings

A new form of protection has been developed by applying certain species of bacterial films to the surface of metals in highly corrosive environments. This process increases the corrosion resistance substantially. Alternatively, antimicrobial-producing biofilms can be used to inhibit mild steel corrosion from sulfate-reducing bacteria.

Controlled Permeability Formwork

Controlled permeability formwork (CPF) is a method of preventing the corrosion of reinforcement by naturally enhancing the durability of the cover during concrete placement. CPF has been used in environments to combat the effects of carbonation, chlorides, frost and abrasion.

Cathodic Protection

Cathodic protection (CP) is a technique to control the corrosion of a metal surface by making that surface the cathode of an electrochemical cell. Cathodic protection systems are most commonly used to protect steel, and pipelines and tanks; steel pier piles, ships, and offshore oil platforms.

Sacrificial Anode Protection

For effective CP, the potential of the steel surface is polarized (pushed) more negative until the metal surface has a uniform potential. With a uniform potential, the driving force for the corrosion reaction is halted. For galvanic CP systems, the anode material corrodes under the influence of the steel, and eventually it must be replaced. The polarization is caused by the current flow from the anode to the cathode, driven by the difference in electrode potential between the anode and the cathode.

Sacrificial anode attached to the hull of a ship

Impressed Current Cathodic Protection

For larger structures, galvanic anodes cannot economically deliver enough current to provide complete protection. Impressed current cathodic protection (ICCP) systems use anodes connected to a DC power source (such as a cathodic protection rectifier). Anodes for ICCP systems are tubular and solid rod shapes of various specialized materials. These include high silicon cast iron, graphite, mixed metal oxide or platinum coated titanium or niobium coated rod and wires.

Anodic Protection

Anodic protection impresses anodic current on the structure to be protected (opposite to the cathodic protection). It is appropriate for metals that exhibit passivity (e.g. stainless steel) and suitably small passive current over a wide range of potentials. It is used in aggressive environments, such as solutions of sulfuric acid.

Rate of Corrosion

These neodymium magnets corroded extremely rapidly after only 5 months of outside exposure

A simple test for measuring corrosion is the weight loss method. The method involves exposing a clean weighed piece of the metal or alloy to the corrosive environment for a specified time followed by cleaning to remove corrosion products and weighing the piece to determine the loss of weight.

The rate of corrosion (R) is calculated as

$$R = \frac{kW}{\rho At}$$

where k is a constant, W is the weight loss of the metal in time t, A is the surface area of the metal exposed, and ρ is the density of the metal (in g/cm^3).

Other common expressions for the corrosion rate is penetration depth and change of mechanical properties.

Economic Impact

The collapsed Silver Bridge, as seen from the Ohio side

In 2002, the US Federal Highway Administration released a study titled "Corrosion Costs and Preventive Strategies in the United States" on the direct costs associated with metallic corrosion in the US industry. In 1998, the total annual direct cost of corrosion in the U.S. was ca. \$276 billion (ca. 3.2% of the US gross domestic product). Broken down into five specific industries, the economic losses are \$22.6 billion in infrastructure; \$17.6 billion in production and manufacturing; \$29.7 billion in transportation; \$20.1 billion in government; and \$47.9 billion in utilities.

Rust is one of the most common causes of bridge accidents. As rust has a much higher volume than the originating mass of iron, its build-up can also cause failure by forcing apart adjacent parts. It was the cause of the collapse of the Mianus river bridge in 1983, when the bearings rusted internally and pushed one corner of the road slab off its support. Three drivers on the roadway at the time died as the slab fell into the river below. The following NTSB investigation showed that a drain in the road had been blocked for road re-surfacing, and had not been unblocked; as a result, runoff water penetrated the support hangers. Rust was also an important factor in the Silver Bridge disaster of 1967 in West Virginia, when a steel suspension bridge collapsed within a minute, killing 46 drivers and passengers on the bridge at the time.

Similarly, corrosion of concrete-covered steel and iron can cause the concrete to spall, creating severe structural problems. It is one of the most common failure modes of reinforced concrete bridges. Measuring instruments based on the half-cell potential can detect the potential corrosion spots before total failure of the concrete structure is reached.

Until 20–30 years ago, galvanized steel pipe was used extensively in the potable water systems for single and multi-family residents as well as commercial and public construction. Today, these systems have long ago consumed the protective zinc and are corroding internally resulting in poor water quality and pipe failures. The economic impact on homeowners, condo dwellers, and the public infrastructure is estimated at 22 billion dollars as the insurance industry braces for a wave of claims due to pipe failures.

Corrosion in Nonmetals

Most ceramic materials are almost entirely immune to corrosion. The strong chemical bonds that hold them together leave very little free chemical energy in the structure; they can be thought of as already corroded. When corrosion does occur, it is almost always a simple dissolution of the material or chemical reaction, rather than an electrochemical process. A common example of corrosion protection in ceramics is the lime added to soda-lime glass to reduce its solubility in water; though it is not nearly as soluble as pure sodium silicate, normal glass does form sub-microscopic flaws when exposed to moisture. Due to its brittleness, such flaws cause a dramatic reduction in the strength of a glass object during its first few hours at room temperature.

Corrosion of Polymers

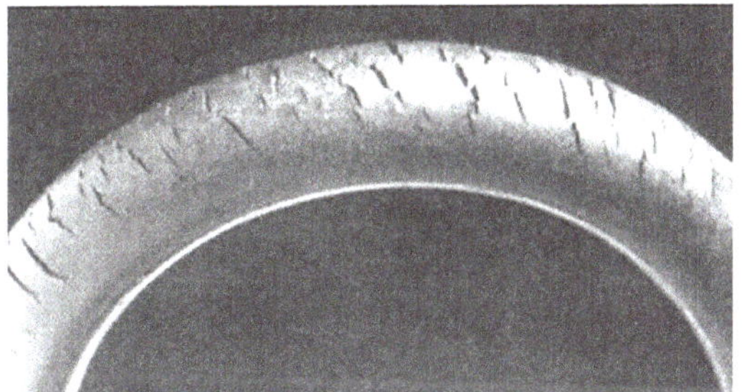

Ozone cracking in natural rubber tubing

Polymer degradation involves several complex and often poorly understood physiochemical processes. These are strikingly different from the other processes discussed here, and so the term "corrosion" is only applied to them in a loose sense of the word. Because of their large molecular weight, very little entropy can be gained by mixing a given mass of polymer with another substance, making them generally quite difficult to dissolve. While dissolution is a problem in some polymer applications, it is relatively simple to design against.

A more common and related problem is "swelling", where small molecules infiltrate the structure, reducing strength and stiffness and causing a volume change. Conversely, many polymers (notably flexible vinyl) are intentionally swelled with plasticizers, which can be leached out of the structure, causing brittleness or other undesirable changes.

The most common form of degradation, however, is a decrease in polymer chain length. Mechanisms which break polymer chains are familiar to biologists because of their effect on DNA: ionizing radiation (most commonly ultraviolet light), free radicals, and oxidizers such as oxygen, ozone, and chlorine. Ozone cracking is a well-known problem affecting natural rubber for example. Plastic additives can slow these process very effectively, and can be as simple as a UV-absorbing pigment (e.g. titanium dioxide or carbon black). Plastic shopping bags often do not include these additives so that they break down more easily as ultrafine particles of litter.

Corrosion of Glasses

Glass corrosion

Glass is characterized by a high degree of corrosion-resistance. Because of its high water-resistance it is often used as primary packaging material in the pharma industry since most medicines are preserved in a watery solution. Besides its water-resistance, glass is also very robust when being exposed to chemically aggressive liquids or gases. While other materials like metal or plastics quickly reach their limits, special glass-types can easily hold up.

Glass disease is the corrosion of silicate glasses in aqueous solutions. It is governed by two mechanisms: diffusion-controlled leaching (ion exchange) and hydrolytic dissolution of the glass network. Both mechanisms strongly depend on the pH of contacting solution: the rate of ion exchange decreases with pH as $10^{-0.5pH}$ whereas the rate of hydrolytic dissolution increases with pH as $10^{0.5pH}$.

Mathematically, corrosion rates of glasses are characterized by normalized corrosion rates of elements NR_i (g/cm²·d) which are determined as the ratio of total amount of released species into the water M_i (g) to the water-contacting surface area S (cm²), time of contact t (days) and weight fraction content of the element in the glass f_i:

$$\mathrm{NR}_i = \frac{M_i}{Sf_it}.$$

The overall corrosion rate is a sum of contributions from both mechanisms (leaching + dissolution) $NR_i = NRx_i + NRh$. Diffusion-controlled leaching (ion exchange) is characteristic of the initial phase of corrosion and involves replacement of alkali ions in the glass by a hydronium (H_3O^+) ion from the solution. It causes an ion-selective depletion of near surface layers of glasses and gives an inverse square root dependence of corrosion rate with exposure time. The diffusion-controlled normalized leaching rate of cations from glasses (g/cm²·d) is given by:

$$NRx_i = 2\rho\sqrt{\frac{D_i}{\pi t}},$$

where t is time, D_i is the i-th cation effective diffusion coefficient (cm²/d), which depends on pH of contacting water as $D_i = D_{io} \cdot 10^{-pH}$, and ρ is the density of the glass (g/cm³).

Glass network dissolution is characteristic of the later phases of corrosion and causes a congruent release of ions into the water solution at a time-independent rate in dilute solutions (g/cm²·d):

$$NRh = \rho r_h ,$$

where r_h is the stationary hydrolysis (dissolution) rate of the glass (cm/d). In closed systems the consumption of protons from the aqueous phase increases the pH and causes a fast transition to hydrolysis. However, a further saturation of solution with silica impedes hydrolysis and causes the glass to return to an ion-exchange, e.g. diffusion-controlled regime of corrosion.

In typical natural conditions normalized corrosion rates of silicate glasses are very low and are of the order of 10^{-7}–10^{-5} g/(cm²·d). The very high durability of silicate glasses in water makes them suitable for hazardous and nuclear waste immobilisation.

Glass Corrosion Tests

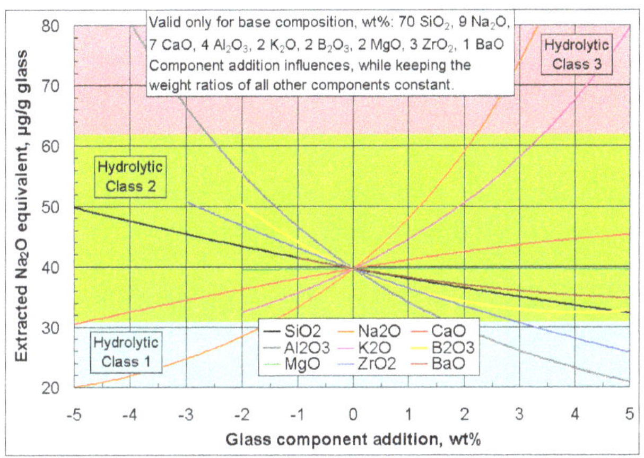

Effect of addition of a certain glass component on the chemical durability against water corrosion of a specific base glass (corrosion test ISO 719).

There exist numerous standardized procedures for measuring the corrosion (also called chemical durability) of glasses in neutral, basic, and acidic environments, under simulated environmental conditions, in simulated body fluid, at high temperature and pressure, and under other conditions.

The standard procedure ISO 719 describes a test of the extraction of water-soluble basic compounds under neutral conditions: 2 g of glass, particle size 300–500 µm, is kept for 60 min in 50 ml de-ionized water of grade 2 at 98 °C; 25 ml of the obtained solution is titrated against 0.01 mol/l HCl solution. The volume of HCl required for neutralization is classified according to the table below.

Amount of 0.01M HCl needed to neutralize extracted basic oxides, ml	Extracted Na_2O equivalent, µg	Hydrolytic class
< 0.1	< 31	1
0.1-0.2	31-62	2
0.2-0.85	62-264	3
0.85-2.0	264-620	4
2.0-3.5	620-1085	5
> 3.5	> 1085	> 5

The standardized test ISO 719 is not suitable for glasses with poor or not extractable alkaline components, but which are still attacked by water, e.g. quartz glass, B_2O_3 glass or P_2O_5 glass.

Usual glasses are differentiated into the following classes:

Hydrolytic class 1 (Type I):

This class, which is also called neutral glass, includes borosilicate glasses (e.g. Duran, Pyrex, Fiolax).

Glass of this class contains essential quantities of boron oxides, aluminium oxides and alkaline earth oxides. Through its composition neutral glass has a high resistance against temperature shocks and the highest hydrolytic resistance. Against acid and neutral solutions it shows high chemical resistance, because of its poor alkali content against alkaline solutions.

Hydrolytic class 2 (Type II):

This class usually contains sodium silicate glasses with a high hydrolytic resistance through surface finishing. Sodium silicate glass is a silicate glass, which contains alkali- and alkaline earth oxide and primarily sodium oxide and Calcium oxide.

Hydrolytic class 3 (Type III):

Glass of the 3rd hydrolytic class usually contains sodium silicate glasses and has a mean hydrolytic resistance, which is two times poorer than of type 1 glasses.

Acid class DIN 12116 and alkali class DIN 52322 (ISO 695) are to be distinguished from the hydrolytic class DIN 12111 (ISO 719).

With the view of achieving higher thermodynamic efficiency, the temperature of coolant leaving the core must be as high as possible. Two challenges must be met while increasing the coolant outlet temperature to higher values:

(i) Maximum allowable temperature coolant temperature has to be fixed taking into consideration of thermo-mechanical resistance of materials. Increasing the coolant outlet temperature beyond this may lead to structural damages to components of reactor or cooling circuits

(ii) The distribution of coolant across the core must ensure minimal occurrences of hot spots.

Coolant Flow Distribution

Coolant flow distribution is an important aspect of cooling circuits of fast reactor. It must be designed to ensure the following:

(i) adequate cooling of each subassembly

(ii) homogeneity of core outlet temperature profile, i.e. the variation in coolant outlet temperatures in radial direction must be minimized

To facilitate the above, subassemblies in the core are classified into different flow zones. This classification is based on the maximum linear power rating of each subassembly, which in turn determines flow rates of coolant in each flow zone. Hence all subassemblies in a specific flow zone are in thermal contact with the same coolant flow rate.

The flow rate in each zone is fixed to ensure that its hottest part is cooled to the desired lower temperature. This is a conservative approach to prevent overheating of subassemblies. Hence there are parts in each flow zone that are slightly overcooled.

Pressure Drop in Core

The pressure drop in the core for flow of sodium determines the energy required for pumping. More important aspect is the maximum pressure drop for the required flow rate of primary sodium. Under steady state or in normal operating conditions, primary sodium flow is maintained by sodium pumps. In the case of transients due to loss of flow due to malfunctioning of pump, the sodium flow is to be maintained by natural convection. The sodium flow by natural convection is due to the reduced density as sodium heats up. Lighter hot sodium moves up the core and leaves at the top while relatively cold sodium enters the core at the bottom. Too a higher-pressure drop would provide higher resistance for flow to be established by natural convection. Hence maximum pressure drop in the core is restricted to 5.4 atm in Prototype Fast Breeder Reactor.

Core Height to Diameter Ratio

Height and diameter of core play an important role in the overall design with respect to safety and economy. With the core of larger length, the cost of subassembly is high apart from increased sodium void coefficient. Higher cost is attributed to difficulty in fabrication of fuel and due to increased subassembly length. With increase in core height, the volume fraction for the fuel is reduced for a fixed pressure drop for coolant flow. Hence both from the perspective of economy and safety, shorter cores are desirable.

But a constraint arises due to the use of shorter cores. The fissile fuel inventory required is larger; hence more number of pins and a larger diameter core are required. The reactor control becomes difficult with too a larger diameter core. Similarly, radial flux flattening for larger diameter core is relatively difficult. Hence, a trade-off needs to be achieved for optimum core height and diameter. In other words, an optimum core height to core diameter ratio needs to be worked out. For most fast reactors, a core of 1 m height is considered to represent this tradeoff satisfactorily, with the height-to-diameter ratio of 0.2 for large diameter cores.

Auxiliary Systems

The ventilation system in a nuclear power plant has important role during both the normal operating condition and in emergency situations. A typical ventilation system consists of ducts that provide flow path for air, and the dampers that control flow rate of air passing through the ducts. Chillers, fan cooler units and air conditioning units cool the air passing through the duct. Fans ensure circulation of air while filters remove impurities in air.

Thus the important components of ventilation system are

(i) Containment Emergency Cooling Systems: This system removes heat from the containment in emergency circumstances during which containment gets heated. The hot air during postulated accident is drawn by a fan and directed towards coils. These coils cool the hot air using water. The cooled air is directed towards the top of the containment to cool the containment.

(ii) Filtration Systems: Fans draw air from the regions that are contaminated. This is passed through a filtration system consisting of PAC filters. 'P' refers to the particulate filters that are responsible for removal of larger particulate matter in the contaminated air. 'A' refers to the absolute filters capable of removing particles with more than 99.5 % efficiency. 'C' refers to the charcoal filters in which the radioactive iodine is adsorbed on the surface of charcoal. The air after passing through PAC filters is directed to exhaust stack.

(iii) Pressure Differential Systems: This system maintains appropriate pressure difference between containment and reactor building & between reactor building and turbine building. The pressure in the reactor building/auxiliary building is higher than that in the containment. This ensures that there are no leakages from containment to the reactor building/auxiliary building. Similarly, pressure in the turbine building is greater than that in the reactor building. These pressures are maintained by controlling the flow rate of air entering and leaving each building through control of appropriate fans.

(iv) Cooling systems comprising air conditioning units and fans

Annulus Gas System

Annulus gas is used in pressurized heavy water reactors (CANDU reactors). The annulus gas passes through the annular region between the coolant and pressure tubes in the calandria. The annulus gas must have low thermal conductivity and be less prone to corrosion. Irradiation of annulus gas should preferably not produce radioactive nucleotides.

The low thermal conductivity for annulus gas is required to minimize the heat loss from the coolant in the coolant tubes to the moderator surrounding the pressure tubes. CO_2 is preferred as annular gas due to its lower thermal conductivity when compared to air and nitrogen, which were used earlier as annulus gas.

Annular gas system is designed to (i) detect the leakage of heavy water from pressure tubes or coolant tubes into the annular region (ii) to provide an assessment of leakage rate of heavy water (iii) to identify the source of leakage and (iv) to locate the source of leakage with minimal exposure of hazardous environment to plant personnel. This system maintains Deuterium level below 0.1 vol % in annulus gas.

The use of CO_2 as annulus gas is not without problems. The issues of chemical reaction and polymer formation still exist with CO_2 as annulus gas. Deuterium dissociates from heavy water due to radiation. Deuterium then enters the annulus gas through diffusion via stainless steel plugs at the end of the tube. Deuterium reacts with CO_2 producing D_2O and CO. The presence of D_2O in annulus gas increases the dew point over a period of time. Measurement of dew point of annulus gas is performed to ascertain the leakage of heavy water. However, heavy water formed due to reaction between deuterium and CO_2 may increase the dew point. This interferes with dew point measurement meant for detection of heavy water leakage. Hence the sensitivity of dew point meters for detection of heavy water leakage is compromised.

The presence of CO in annulus gas due to reaction of CO_2 with deuterium leads to the formation of several organic compounds that constrict the flow passage for annulus gas and hence reduce the flow of annulus gas. This is another factor that compromises the detection of heavy water leakages.

Transmutation reactions with carbon represent another challenge. ^{14}C is the major isotope produced which requires radiological protection to be ensured to personnel during maintenances. Gamma radiation monitors are installed to alert plant personnel in the event of high radiation levels.

Process Water Systems

Process water systems of PHWR comprise two systems: (i) Active Process Water System and (ii) Process Water Cooling System.

The Active Process Water System (APWS) acts as a barrier between radiation producing heat sources and the Process Water Cooling System (PWCS). APWS is a part of secondary and intermediate cooling system related to the plant's safety.

Among several components and systems cooled by APWS, the important systems are

(i) moderator system

(ii) safety related equipment in reactor and service building

(iii) pumps for shutdown cooling system, moderator pumps, ECCS heat exchangers, containment coolers etc.

The Process Water Cooling System (PWCS) is a system that is directly connected to Ultimate Heat Sink (UHS). A large source of water such as lake, river or sea may be used as USH. In case, the plant is located away from a large water source, air may be used as ultimate heat sink to which the water would reject heat in a cooling tower.

The primary objective of PWCS is to remove heat from APWS through process water heat exchangers. Apart from this, cooling to diesel generators, compressors etc. are also provided by this system.

Fire Water Systems

This system is designed with the primary objective of providing water meant for fire fighting. However, if required, the system may also be used as backup cooling for minimizing the effects of certain design basis events, which include station blackout, and failure of Active Process Water System or Process Water Cooling System.

To facilitate the use of fire water system for purposes other than fire fighting stated above, this system is provided with dedicated pumps driven by diesel generators. During station blackout, these pumps supply water to steam generators.

The overall function of secondary system is to ensure regulated supply of steam to various turbine units, condense the spent steam from turbines and process the condensate to a condition suitable for re-use in the steam generator as the secondary coolant.

The secondary system of a typical Pressurized Water Reactor comprises two important subsystems: (a) steam system & (b) condensate/feed water system. The secondary system of PWR is located outside the containment and does not contain any radioactive material. The steam generated in a Boiling Water Reactor is slightly radioactive with N-16 being the major isotope with a half-life of 7 seconds. Hence the secondary system of BWR is shielded to protect plant personnel from exposure to radiation during the normal operations. The short half-life of N-16 makes the secondary system accessible to maintenance within a short duration of reactor shutdown.

Layout of Secondary System

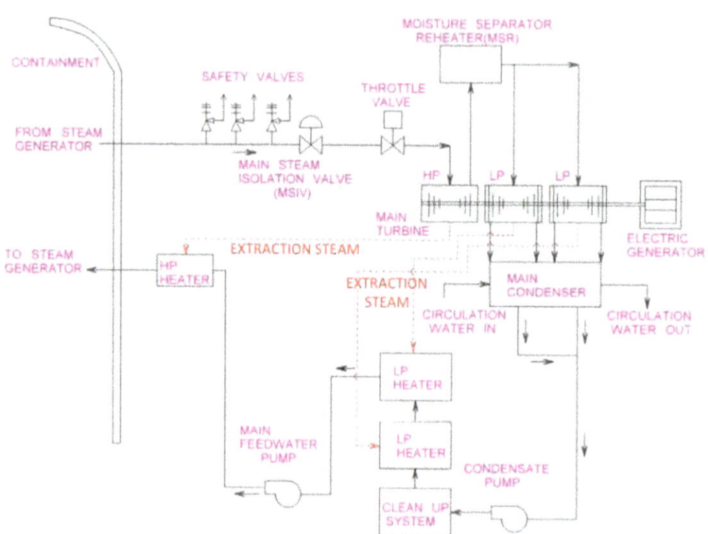

Schematic diagram of secondary system of PWR (Redrawn from Ref.)

The outlet of steam generator represents the beginning of secondary system. Multiple parallel lines are used to supply steam from (steam) generator to the high-pressure turbine. The steam leaving the high-pressure turbine enters the moisture separator and reheater, before being directed to the low-pressure turbines. The exhaust steam from low-pressure turbine is condensed using the cooling water circulated in the condenser. The condensate is pumped through clean up system and low-pressure heaters using the condensate pump. The main feedwater pump is used to pump water obtained from low-pressure heaters to high-pressure heater, from which water is supplied as secondary coolant to the steam generator. A schematic layout of secondary system is shown in Figure.

Important Components of Secondary System

From the layout of secondary system shown in Figure, the following may be identified as important components of secondary system:

Valves

A number of valves including safety valves, isolation valve and throttle valve are located in the pipelines connecting steam generator with high-pressure turbine. The isolation valve located outside the containment is used to isolate the reactor from turbine section, in case of a requirement. Safety valves ensure that the pressure in the pipelines carrying steam from generator to high-pressure turbine is within permissible limits, by bleeding the excess steam responsible for pressure increase. Throttle valve is used to regulate the flow rate of steam entering the high-pressure turbine.

Turbines

The conversion of thermal energy of high-pressure steam to mechanical energy is accomplished in turbines. These turbines drive the generator connected to them via a common shaft rotated by steam impinging on turbine blades.

In a typical PWR, the steam entering the high-pressure turbine is saturated at about 6.68 MPa with moisture content of about 0.25 %. Here it is important to note the difference in the quality of steam produced in a thermal power plant using fossil fuels and that produced in steam generator of nuclear reactor. In fossil-fuel operated power plants, steam is available at very high pressures (16-17 MPa) and temperature (540 C). Supercritical boilers produce steam with much higher pressures (25 MPa). The lower steam pressure results in its higher specific volume. This means that 1 kg of steam obtained from SG of a PWR occupies more volume than 1 kg of steam of obtained from boiler of a typical thermal power plant. The total energy content of a wet steam is lower than that of dry steam. Hence to perform an estimated work, higher mass of steam is required if it is wet. The wet nature of steam from PWR combined with lower pressure leads to higher steam requirement, both in terms of mass and volume, when compared to that required for a typical thermal power plant. For instance, a 1000 MWe thermal power plant requires 800 kg/s of steam while that of a 1000 MWe PWR would require 1700 kg/s of steam.

The higher volumetric flow rate of steam in PWR requires turbines with blades of larger diameter and length. To maintain the blade tip speed (pDN) within limits, larger diameter blades must be run at lower speeds. Hence for a fixed frequency, say 50 Hz, the turbines in nuclear power plants are run at speed, which is 50 % of turbine speed in thermal power plants.

The high-pressure turbine used in PWR is normally of double-flow type. Depending on the number of exhausts in turbines, they are classified as single-flow type and double-flow type. In single-flow type, steam enters at one of the turbine, expands by performing work and leaves at the other end of the turbine. With the use of large volume of steam and high pressure drop across the blades incurred, there exists an imbalance of axial forces on the rotor. This can be circumvented by supplying steam at the centre of turbine and allowing the same to expand in both directions with exhaust located at both the ends of turbine. The axial force exerted on the rotor at one end is balanced by those exerted on the rotor at the other end. Such turbines with steam inlet at the centre and one steam exhaust at each end are called double-flow turbines.

Steam volume at the exhaust of high-pressure turbine is much higher than that at the inlet. Hence two, double-flow low-pressure turbines are required. The low-pressure turbines directly deliver exhaust steam to a condenser maintained at negative pressure. The use of negative pressure in condenser permits removal of larger amount of heat from steam in low-pressure turbines.

Example - 1: Steam at the rate of 500 kg/s boiler is produced in a boiler. The analysis of steam showed the presence of liquid water to the extent of 0.5 %. Determine the mass flow rate of dry steam.

Solution:

Let m_w, m_s and m_{ws} be the mass flow rates of water, dry steam and wet steam respectively.

From the mass balance, $m_{ws} = m_w + m_s$

% wet steam = m_w*100/m_{ws} = m_w*100/(m_w+m_s)

Solving the above equation by setting %wet steam as 5 %, mass flow rate of dry steam is found to be 497.5 kg/s

Example – 2: The specific enthalpy of dry steam and saturated water at 6.8 MPa and 285 C (saturation pressure and temperature) are 2773 kJ/kg and 1262 kJ/kg respectively. If the dryness fraction of steam is 99 %, determine the specific enthalpy of wet steam.

Solution:

Dryness fraction, $x_d = 0.99$

Specific enthalpy of dry steam, h_g = 2773 kJ/kg

Specific enthalpy of saturated water, h_l=1262 kJ/kg

Specific enthalpy of wet steam, $h_{ws} = x_d h_g + (1-x_d) h_l$ = 0.99*2773+0.01*1262 = 2757.89 kJ/kg

Therefore, specific enthalpy of steam with 99 % dryness under these conditions is 2757.89 kJ/kg.

Moisture Separator

The wetness of steam increases as it is expanded in high-pressure turbine. The presence of moisture in the steam leads to corrosion of turbine blades, apart from reducing the energy of steam that can be recovered (lower turbine efficiency). Hence the moisture content of steam or in other words, the wetness fraction of steam must not be greater than 10 % during its transit in any of the turbines (high-pressure or low- pressure) to avoid damage to turbine blades.

The spent steam from high-pressure turbine is directed to the moisture separators and then to reheaters. In some designs, moisture separator and reheater are integrated into a single unit named moisture separator reheater (MSR). Stand-alone moisture separators work by imparting centrifugal motion to steam-water mixture through use of vanes. This causes moisture to be disengaged from the wet steam. A schematic diagram of stand-alone moisture separator is shown in Figure.

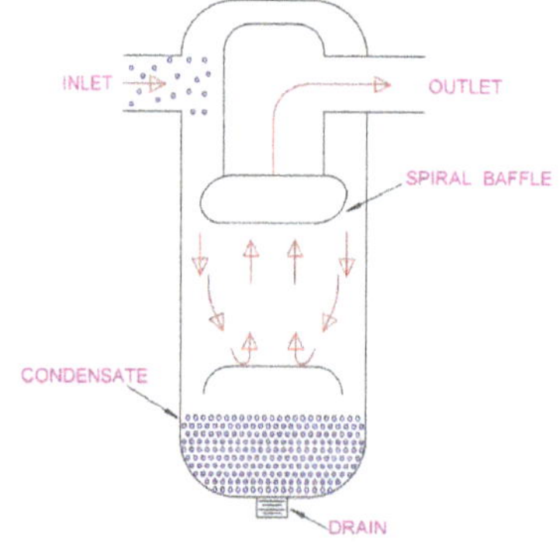

Schematic diagram of stand-alone moisture separator (Redrawn from Ref.)

In MSR systems, moisture is separated from the steam by directing steam-water mixture through chevron type plates that cause steam to change flow direction rapidly. A schematic diagram of moisture separator in integrated MSR systems is shown in Figure.

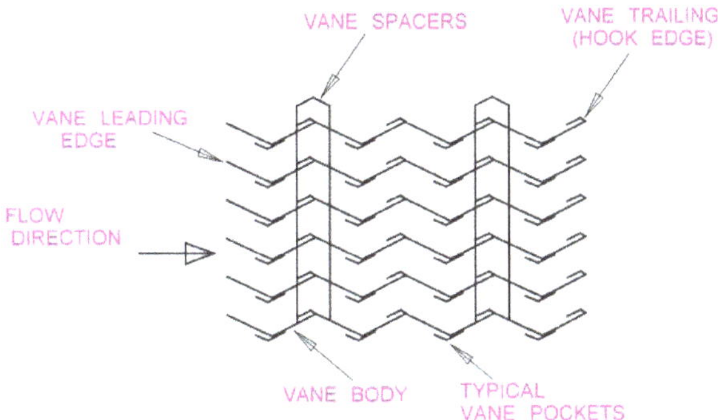

Schematic diagram of moisture separator in integrated MSR systems (Redrawn from Ref.)

Reheater

We have realized the importance of maintaining a lower wetness fraction of steam to prevent corrosion of turbine blades as well as to achieve higher turbine efficiency. Though moisture separators remove majority of moisture, reheaters are required to improve dryness fraction or quality of steam.

Reheaters utilize finned tubes as heat transfer surface. The hot fluid for reheaters is the high-pressure steam from the steam generator and passes inside the finned tubes. In PWR, high-pressure steam at 5.5 MPa is used for this purpose. The exhaust steam after passing through the moisture separator flows outside the tubes. Though both the high-pressure steam used for heating and the exhaust steam are saturated, a temperature difference of about 100 exists between the two streams due to difference in their pressures. The exhaust steam gets superheated upon thermal contact with the high-pressure steam thereby becoming dry vapors. The heat transfer coefficient for heat transfer involving dry gas or vapor is low. Hence, finned tubes are used to increase the heat transfer area to compensate for lower heat transfer coefficient. The temperature of exhaust steam after reheating reaches within 20°of the hot stream used for reheating and is suitable for feeding to low-pressure turbines. The high-pressure steam used for reheating condenses inside the tubes and is supplied to high-pressure (feedwater) heater as the hot fluid. The schematic diagrams of MSR and stand-alone reheater are shown in Figures.

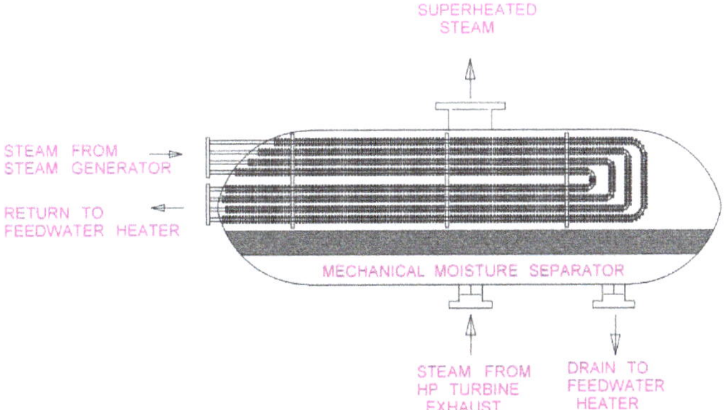

Schematic diagram of moisture separator reheater (Modified from Ref.)

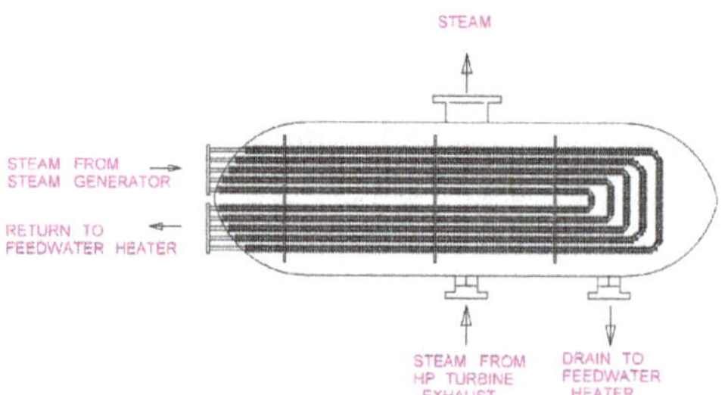

Schematic diagram of stand-alone reheater

Condensate/feed water system

Let us recall the components of 'condensate/feedwater system' from the layout of secondary systeme. The important components are (i) condenser; (ii) cooling tower; (iii) condensate pump; (iv) cleanup system; (v) low pressure heaters; (vi) main feedwater pump and (vii) high pressure heater.

Condenser

The condensate/feedwater system begins with the condensation of exhaust steam from low-pressure turbines in a condenser operating under vacuum. Condenser is a shell and tube heat exchanger with steam condensing on the outer surface of tubes, with coolant supplied through the tubes. The schematic diagram of a typical condenser is shown in Figure.

The important components of the condenser are

(i) tube bundle

(ii) tube sheet

(iii) inlet nozzle for steam

(iv) hot well

(v) nozzle for condensate outlet

(vi) nozzle for coolant inlet

(vii) nozzle for coolant outlet

(viii) pass partition

(ix) nozzle for ejector

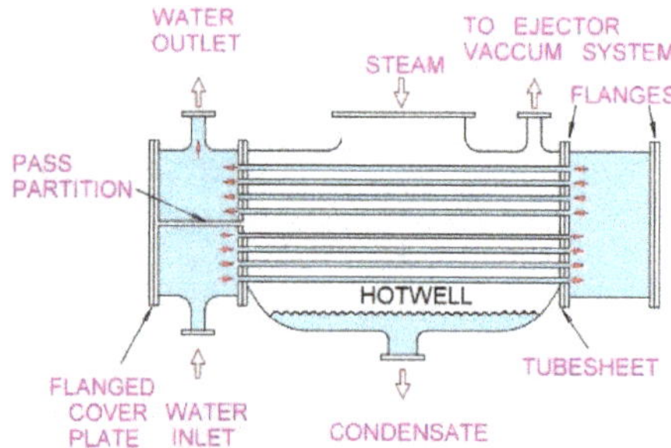

Schematic diagram of shell & tube condenser

The condenser shown in Figure is a shell and tube heat exchanger, with two passes on the tube side and one pass on the shell side. Cooling water enters the condenser through inlet nozzle located near the bottom. The pass partition confines the cross sectional area available for coolant flow to 50 % of total cross sectional area of all the tubes. This arrangement ensures that the velocity of coolant flow inside the tubes is high enough to promote turbulence and enhance the rate of heat transfer. The tubes are arranged in a triangular pitch as shown in Figure and are held together at both ends using tube sheets as shown in Figure. The centre-to-centre distance between two tubes is called pitch, which is maintained between 1.25-1.5 times the outer diameter of the tube. With the triangular pitch, more tubes can be accommodated per unit heat exchanger volume. In other words, higher heat transfer area per unit heat exchanger volume can be obtained with triangular pitch than that obtainable from a square pitch. However, the accessibility of outer surface of the tubes for cleaning becomes difficult. Tube sheets also provide physical barrier to prevent mixing of shell-side and tube-side fluids. Steam enters the condenser through a nozzle located at the top. The nozzle for steam inlet is larger in diameter compared to the nozzles for coolant inlet and outlet. The density of steam is lower than that of water. Hence larger diameter nozzle is required for steam inlet.

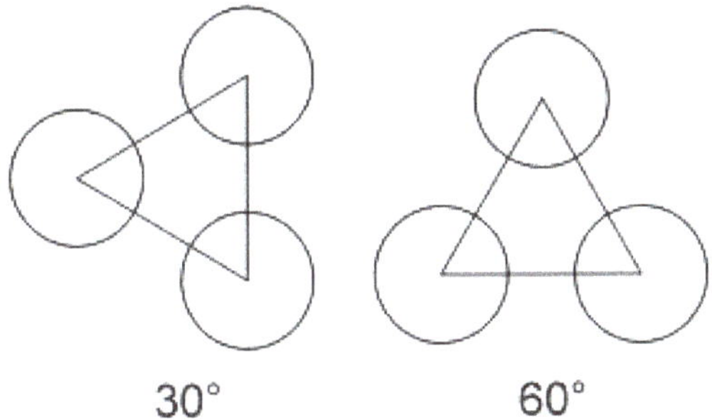

Arrangement of tubes in triangular pitch

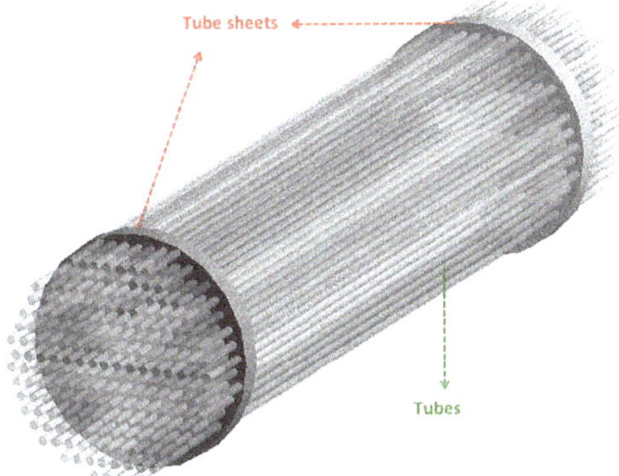

Tube sheets holding the tubes together at both ends

Steam is condensed on outer side of the tubes by thermal contact with cooling water flowing inside the tubes. The tubes are generally made of stainless steel to overcome corrosion. The condensate water is collected in the hot well from which the same is pumped using a condensate pump for further chemical treatment and heating. The pressure inside the condenser is maintained below atmospheric by steam ejectors that remove non-condensibles. When steam condenses, its specific volume is reduced. This creates vacuum that draws the exhaust steam from the low-pressure turbines.

Lower temperature of cooling water is advantageous to achieve higher efficiency of power plant.

The temperature of cooling water after passing through the condenser is increased by around 10-15°C. When the cooling water for condenser is taken from a perennial water source like sea or river, the same can be dumped in to the source. The large volume of water in the source ensures rapid equilibration of temperature to ambient level. In case of power plants that are located away from such perennial water sources, cooling towers are employed to bring down the coolant temperature by evaporative cooling.

Cooling Tower

Cooling towers are available in several designs, configurations and sizes depending on the cooling requirements. Among different types of cooling towers, natural draft and induced draft cooling towers are widely used in industries.

Natural draft cooling towers rely on the difference in densities of hot and cold air for flow of air through the tower. Cold air enters the tower at the bottom and is heated due to direct contact with water to be cooled, which is supplied in the form of fine sprays. The hot air, being denser rises upwards. This creates vacuum inside the tower. The relatively cool, ambient air at atmospheric pressure enters the tower due to this vaccum; gets heated upon direct contact with sprayed water and leaves at the top. The major advantage of natural draft cooling towers are the reduction in operating cost due to absence of mechanical equipment to induce the movement of air.

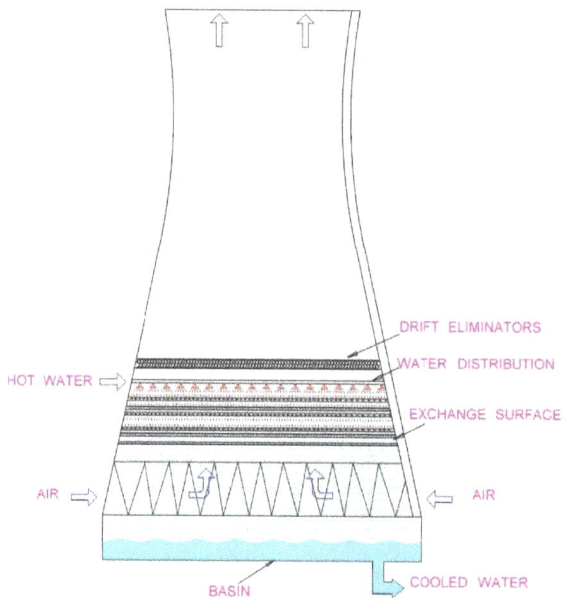

Hyperbolic natural draft towers used in power plants

Hyperbolic natural draft towers shown in Figure are commonly used in power plants. The shell of these cooling towers is hyperbolic in shape and hence called hyperbolic natural draft towers. This geometry facilitates the rapid movement of hot air towards the upper portions of the tower. These towers are fairly large in diameter as well as in height. The cooling water from the condenser is sprayed across the cross section of the tower at the bottom. The air entering the tower is unsaturated. In simple language, the water vapor content of unsaturated air is lower than the maximum water vapor carrying capacity of air at a given temperature. For instance, we often come across weather reports where temperature and relative humidity are mentioned. Air with relative humidity less than 100 % is unsaturated. When such unsaturated air comes into contact with water, the difference between partial pressure of water vapor in air and the vapor pressure of water at its temperature acts as the driving force for evaporation of water. Since the temperature of air is lower than that of water, the heat required for evaporation of water must be utilized from water itself. This causes cooling of water and hence the term 'evaporative cooling' is used. The cooled water is collected at the bottom of tower in a tank and is pumped back to the condenser as cooling water.

Induced draft cooling towers use fan located at top of the cooling tower, to force the flow of air across the cooling tower coming into direct contact with water sprayed as droplets. The velocity of air at discharge is 3-4 times greater than air velocity at the tower inlet. This ensures that recirculation of exhaust air towards the inlet is not seriously high. These towers can be built with a wide range of sizes and are not as tall as the hyperbolic natural draft towers. The major disadvantage of induced draft cooling towers over the hyperbolic natural draft towers is the energy required for operating the fans.

Other Components

Having seen the major components of condensate/feedwater system, let us discuss the role of other components.

The condensate pump takes inlet from the hot well of condenser and increases its pressure and pumps the condensate through clean up system and low-pressure heaters. Clean up system essentially removes the impurities that can form hard scales inside the tubes of steam generators. Such scales provide additional resistance to heat transfer, causing reduction in rate of heat transfer between primary and secondary coolants leading to reduction in steam generation. The condensate then passes through low-pressure heaters where the condensate is heated using the 'extraction steam' obtained from low-pressure turbines. The condensate is then supplied to the suction side of main feedwater pump, which increases the pressure high enough to permit entry into the steam generator. Before water enters the steam generators, it is heated in high-pressure heater using the 'extraction steam' obtained from high-pressure turbines. The heating of feedwater supplied to steam generator using the extraction steam increases the overall plant efficiency.

Comparison of Operating Conditions of Thermal and Nuclear Power Plants

We shall compare some of the operation conditions of thermal and nuclear power plants. The focus here is on the generation of electricity from steam or the thermodynamic cycles, rather than the comparison of modes of steam generation and the thermal hydraulic characteristics. Table is a compilation of characteristics of steam cycles of some water/heavy water colled nuclear power plants and thermal power plants.

Table: Characteristics of steam cycles of some water/heavy water colled nuclear power plants and thermal power plants

Characteristics	Pressurized water reactor (Westinghouse Co.)	Boiling water reactor (General Electric Co.)	Pressurized heavy water reactor (Atomic Energy of Canada Ltd.)	Thermal power plant
Steam pressure (MPa)	5.7 MPa	7 MPa	4.7 MPa	
Steam temperature (°C)	273 °C	288 °C	260 °C	525 °C
Steam quality	Wet steam	Wet steam	Wet steam	Superheated steam
Efficiency (%)	33.5 %	32.9 %	29.3 %	
Power (MWe)	1148	1178	638	

The following observations can be made from the above table:

- Steam temperature: Higher for thermal power plant compared to nuclear power plant
- Steam quality: Higher steam quality (superheated steam) in thermal power plant compared to nuclear power plant
- Efficiency: Owing to higher steam temperature and better steam quality, thermal power plants are more efficient compared to nuclear power plant
- Steam requirement: As a result of lower steam quality and steam temperature in nuclear power plant, higher mass flow rate of steam required per MW of electricity generated.

References

- Churchill, Stuart W.; Chu, Humbert H.S. (November 1975). "Correlating equations for laminar and turbulent free convection from a vertical plate". International Journal of Heat and Mass Transfer. 18 (11): 1323–1329. doi:10.1016/0017-9310(75)90243-4. Retrieved 18 September 2015
- Sukhatme, S. P. (2005). A Textbook on Heat Transfer (Fourth ed.). Universities Press. pp. 257–258. ISBN 8173715440
- Daniel Robles. "Potable Water Pipe Condition Assessment For a High Rise Condominium in The Pacific Northwest". GSG Group, Inc. Dan Robles, PE. Retrieved 10 December 2012
- Chiavazzo, Eliodoro; Ventola, Luigi; Calignano, Flaviana; Manfredi, Diego; Asinari, Pietro (2014). "A sensor for direct measurement of small convective heat fluxes: Validation and application to micro-structured surfaces". Experimental Thermal and Fluid Science. 55. doi:10.1016/j.expthermflusci.2014.02.010
- James R. Welty; Charles E. Wicks; Robert E. Wilson; Gregory L. Rorrer (2007). Fundamentals of Momentum, Heat and Mass transfer (5th edition). John Wiley and Sons. ISBN 978-0470128688
- Vapor Hydration Testing (VHT) Archived December 14, 2007, at the Wayback Machine.. Vscht.cz. Retrieved on 2012-07-15
- R. Zuo; D. Örnek; B.C. Syrett; R.M. Green; C.-H. Hsu; F.B. Mansfeld; T.K. Wood (2004). "Inhibiting mild steel corrosion from sulfate-reducing bacteria using antimicrobial-producing biofilms in Three-Mile-Island process water". Appl. Microbiol. Biotechnol. 64 (2): 275–283. doi:10.1007/s00253-003-1403-7
- JE Breakell, M Siegwart, K Foster, D Marshall, M Hodgson, R Cottis, S Lyon. Management of Accelerated Low Water Corrosion in Steel Maritime Structures, Volume 634 of CIRIA Series, 2005, ISBN 0-86017-634-7
- International Organization for Standardization, Procedure 719 (1985). Iso.org (2011-01-21). Retrieved on 2012-07-15

4

Types of Thermal Reactors

Boiling water reactor is used for the production of electrical power. It is a common type of nuclear reactor that is used in generating electricity. The other type of thermal reactor discussed in this chapter is gas-cooled reactor. The aspects elucidated in this chapter are of vital importance, and provide a better understanding of nuclear science and technology.

Water Reactor

As the name of the reactor implies, the pressurized water reactor uses light water at high pressure as the coolant. Typically the operating pressure is about155 bar (~ 153 atm). The pressurized water reactor utilizes enriched uranium (with U-235 content about 3 %) as the fuel and light water as both moderator and coolant. Compared to Heavy water, the moderating capability of light water is less. Hence enriched uranium is required to sustain the fission when light water is used as moderator. A schematic diagram of a power plant utilizing Pressurized Water Reactor is shown in Figure.

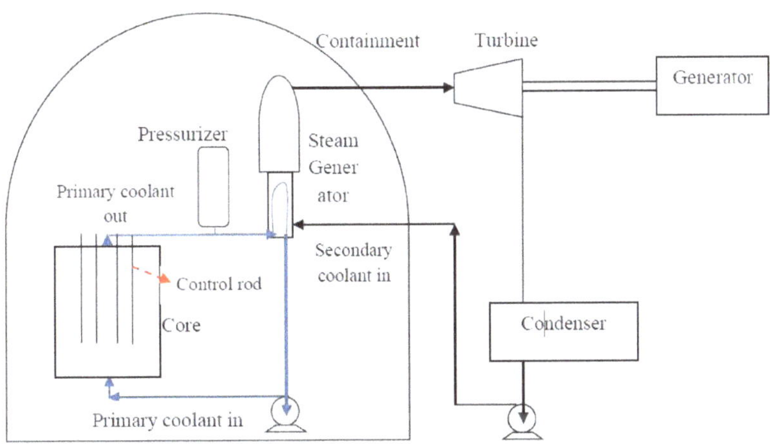

Schematic diagram of Pressurized Water Reactor

The important components of a PWR are (i) A core that contains the fuel (ii) Steam generators (iii) Pressurizer. These components are housed inside the containment (concrete dome that is visible in the plant site). Other components required for power generation like turbine, generator and condenser are located outside the containment.

Two coolant cycles (primary and secondary coolant cycles, with water used as coolant for both cycles) are used in a pressurized water reactor. The primary coolant is circulated through the core in order to remove the heat generated by nuclear fission. The primary coolant maintained at a pressure of ~152 bar, enters at about 288 °C and leaves at about 324 °C depending upon the core configuration. The high pressure ensures that water remains in liquid state even at high temperatures (boiling point of water at 152 bar is 343 °C). The pressure in the primary loop is maintained by a pressurizer.

In the steam generator, primary coolant comes into thermal contact with the secondary coolant. Primary coolant passes through large number of tubes in steam generator, while the secondary coolant (inlet temperature ~ 227 °C) occupies the region between the tubes. This facilitates thermal contact between primary and secondary coolants. The pressure in the steam generator (69 bar) is lower than the reactor operating pressure. Accordingly secondary coolant (also water) boils at a lower temperature (285 °C) than that of the primary coolant. Hence, heat transferred from the primary coolant to secondary coolant (flowing in the annular region between the tubes and shell of steam generator) results in the vaporization of secondary coolant. This results in the generation of steam in the steam generator. Steam, thus produced is used to run the turbine for power generation. The spent steam is condensed in a condenser and returned to the steam generator as secondary coolant along with make-up water. The primary coolant, after transferring heat to the secondary coolant, returns to the core to remove the heat generated during nuclear fission. Thus, the Pressurized Water Reactor follows two-coolant cycle or indirect cycle.

The primary coolant loop contains Reactor Pressure Vessel (RPV), Pressurizer, Steam Generator and Coolant pump. Reactor Pressure Vessel (RPV) consists of core, water and control rods. The reactor vessel has a cylindrical body with the covers (head and bottom) being hemispherical in shape. The top cover (head) is removable. This permits refuelling of the reactor. The reactor vessel is provided with inlet nozzle (cold leg) and outlet nozzle (hot leg) for each primary coolant loop. The material of construction of the reactor vessel is manganese molybdenum steel. The parts of reactor vessel in direct contact with coolant are clad with stainless steel. This increases the resistance of material to corrosion. The core is enclosed in a barrel-shaped structure called 'core barrel'. Fuel assemblies, moderator and coolant are housed in the core. The coolant enters the reactor vessel through the inlet nozzle and impinges on the core barrel. The coolant is directed in the space between reactor vessel and core barrel and reaches the bottom of reactor vessel. The coolant changes its path and moves around and through the fuel assemblies, thereby removing the fission heat. The heated water leaves vessel through outlet nozzle (hot leg) and enters steam generator.

Pressurized Water Reactor Core

Let us look at the details of core of a typical PWR designed by Westinghouse Electric Company, USA. Fuel (UO_2 in this case) is converted to a pellet (8.2 mm diameter and 13.5 mm long) and then clad in tubes of Zircaloy (98 % Zr, 1.5 % Sn, rest other metals). The fuel rod is about 3.7 m long and 1 cm in diameter and can accommodate about 271 fuel pellets. A small space is provided between the pellets and the fuel rod to account for the expansion of the fuel pellets and to accommodate gaseous fission products like Xenon and Krypton. An array of 17 x 17 fuel rods constitutes a fuel bundle or assembly. Allowing few spaces in the array for control rods, approximately 263 rods comprise an assembly. About such 193 fuel assemblies make up the core, resulting in 50,952 fuel rods and about 13,807,992 fuel pellets. Control material (silver-indium-cadmium alloy) is used in the form of rod. These rods are inserted into the core from the top to control the reactor power. The steps involved in conversion of a fuel pellet to fuel assembly are shown in Figure.

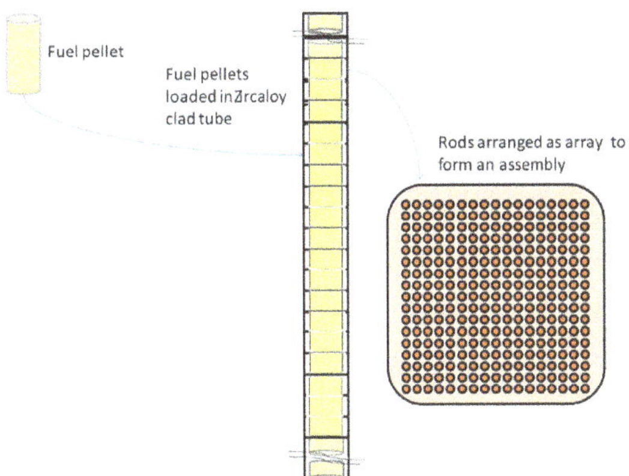

Fuel pellet to fuel assembly

Pressurizer

Pressurizer is a vertical, cylindrical vessel whose bottom is connected to the reactor coolant system (primary coolant) by a single piping. The pressurizer contains stagnant water and steam. The pressure is regulated by varying the temperature of water. When the pressure in the primary coolant loop decreases, electrical heaters placed in pressurizers are activated, heating the water. As water is heated in closed system both temperature and pressure increase (due to generation of steam). In case of higher pressure in the primary loop, a spray located at the top of the pressurizer is activated resulting in condensation of steam (by water spray) and subsequent reduction in pressure.

A schematic diagram of pressurizer is shown below:

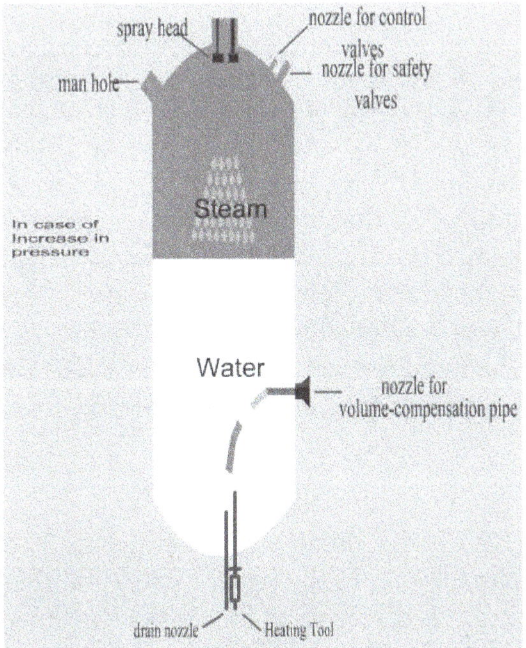

Schematic diagram of Pressurizer (Redrawn from Ref.)

Among several support systems for the primary coolant, one of the important systems is Chemical and Volume Control System (CVCS). This system is designed to (i) maintain the purity of primary coolant with the help of filters (ii) maintain the boron concentration at desired levels by addition of boron or by replacing a part of borated water with water (dilution) and (iii) to maintain water level in pressurizer at desired level. For the purpose of maintaining purity of primary coolant, a small amount of water is continuously passed through this system. The details of steam-water circuit comprising steam generators, turbines, condensers and feed water heaters.

Boiling Water Reactor

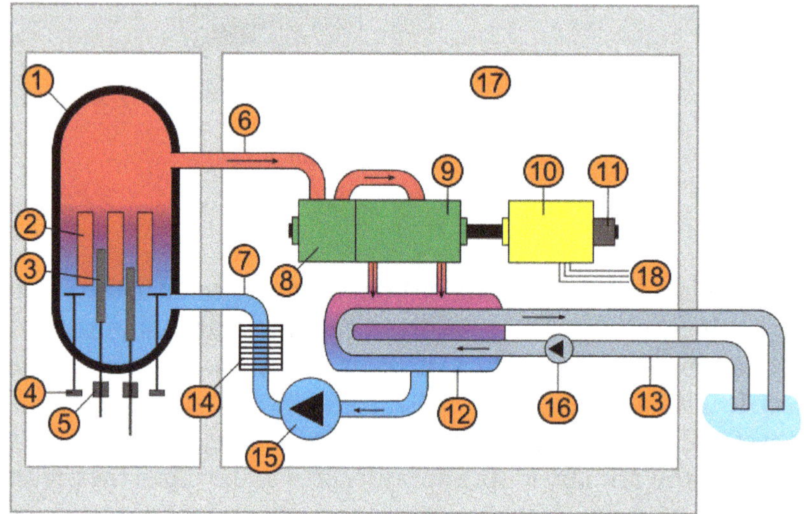

Schematic diagram of a boiling water reactor (BWR):
1. Reactor pressure vessel 2. Nuclear fuel element 3. Control rods 4. Recirculation pumps 5. Control rod drives 6. Steam 7. Feedwater 8. High pressure turbine 9. Low pressure turbine 10. Generator 11. Exciter 12. Condenser 13. Coolant 14. Pre-heater 15. Feedwater pump 16. Cold water pump 17. Concrete enclosure 18. Connection to electricity grid

The boiling water reactor (BWR) is a type of light water nuclear reactor used for the generation of electrical power. It is the second most common type of electricity-generating nuclear reactor after the pressurized water reactor (PWR), also a type of light water nuclear reactor. The main difference between a BWR and PWR is that in a BWR, the reactor core heats water, which turns to steam and then drives a steam turbine. In a PWR, the reactor core heats water, which does not boil. This hot water then exchanges heat with a lower pressure water system, which turns to steam and drives the turbine. The BWR was developed by the Idaho National Laboratory and General Electric (GE) in the mid-1950s. The main present manufacturer is GE Hitachi Nuclear Energy, which specializes in the design and construction of this type of reactor.

Overview

The *boiling water reactor* (BWR) uses demineralized water as a coolant and neutron moderator. Heat is produced by nuclear fission in the reactor core, and this causes the cooling water to boil, producing steam. The steam is directly used to drive a turbine, after which it is cooled in a condenser and converted back to liquid water. This water is then returned to the reactor core, completing the loop. The cooling water is maintained at about 75 atm (7.6 MPa, 1000–1100 psi) so that it boils in the core at about 285 °C (550 °F). In comparison, there is no significant boiling allowed in a pressurized water reactor (PWR) because of the high pressure maintained in its primary loop—approximately 158 atm (16 MPa, 2300 psi). The core damage frequency of the reactor was estimated to be between 10^{-4} and 10^{-7} (i.e., one core damage accident per every 10,000 to 10,000,000 reactor years).

Components

Condensate and Feedwater

Steam exiting the turbine flows into condensers located underneath the low-pressure turbines, where the steam is cooled and returned to the liquid state (condensate). The condensate is then pumped through feedwater heaters that raise its temperature using extraction steam from various turbine stages. Feedwater from the feedwater heaters enters the reactor pressure vessel (RPV) through nozzles high on the vessel, well above the top of the nuclear fuel assemblies (these nuclear fuel assemblies constitute the "core") but below the water level.

The feedwater enters into the downcomer or annulus region and combines with water exiting the moisture separators. The feedwater subcools the saturated water from the moisture separators. This water now flows down the downcomer or annulus region, which is separated from the core by a tall shroud. The water then goes through either jet pumps or internal recirculation pumps that provide additional pumping power (hydraulic head). The water now makes a 180-degree turn and moves up through the lower core plate into the nuclear core, where the fuel elements heat the water. Water exiting the fuel channels at the top guide is saturated with a steam quality of about 15%. Typical core flow may be 45,000,000 kg/h (100,000,000 lb/h) with 6,500,000 kg/h (14,500,000 lb/h) steam flow. However, core-average void fraction is a significantly higher fraction (~40%). These sort of values may be found in each plant's publicly available Technical Specifications, Final Safety Analysis Report, or Core Operating Limits Report.

The heating from the core creates a thermal head that assists the recirculation pumps in recirculating the water inside of the RPV. A BWR can be designed with no recirculation pumps and rely entirely on the thermal head to recirculate the water inside of the RPV. The forced recirculation head from the recirculation pumps is very useful in controlling power, however, and allows achieving higher power levels that would not otherwise be possible. The thermal power level is easily varied by simply increasing or decreasing the forced recirculation flow through the recirculation pumps.

The two-phase fluid (water and steam) above the core enters the riser area, which is the upper region contained inside of the shroud. The height of this region may be increased to increase the thermal natural recirculation pumping head. At the top of the riser area is the moisture separator. By swirling the two-phase flow in cyclone separators, the steam is separated and rises upwards towards the steam dryer while the water remains behind and flows horizontally out into the downcomer or annulus region. In the downcomer or annulus region, it combines with the feedwater flow and the cycle repeats.

The saturated steam that rises above the separator is dried by a chevron dryer structure. The "wet" steam goes through a tortuous path where the water droplets are slowed down and directed out into the downcomer or annulus region. The "dry" steam then exits the RPV through four main steam lines and goes to the turbine.

Control Systems

Reactor power is controlled via two methods: by inserting or withdrawing control blades and by changing the water flow through the reactor core.

Positioning (withdrawing or inserting) control rods is the normal method for controlling power when starting up a BWR. As control rods are withdrawn, neutron absorption decreases in the control material and increases in the fuel, so reactor power increases. As control rods are inserted, neutron absorption increases in the control material and decreases in the fuel, so reactor power decreases. Differently from the PWR, in a BWR the control rods (boron carbide plates) are inserted from below to give a more homogeneous distribution of the power: in the upper side the density of the water is lower due to vapour formation, making the neutron moderation less efficient and the fission probability lower. In normal operation, the control rods are only used to keep a homogeneous power distribution in the reactor and compensate the consumption of the fuel, while the power is controlled through the water flow. Some early BWRs and the proposed ESBWR (Economic Simplified BWR made by General Electric Hitachi) designs use only natural circulation with control rod positioning to control power from zero to 100% because they do not have reactor recirculation systems.

Changing (increasing or decreasing) the flow of water through the core is the normal and convenient method for controlling power from approximately 30% to 100% reactor power. When operating on the so-called "100% rod line," power may be varied from approximately 30% to 100% of rated power by changing the reactor recirculation system flow by varying the speed of the recirculation pumps or modulating flow control valves. As flow of water through the core is increased, steam bubbles ("voids") are more quickly removed from the core, the amount of liquid water in the core increases, neutron moderation increases, more neutrons are slowed down to be absorbed by the fuel, and reactor power increases. As flow of water through the core is decreased, steam voids remain longer in the core, the amount of liquid water in the core decreases, neutron moderation decreases, fewer neutrons are slowed down to be absorbed by the fuel, and reactor power decreases.

Reactor pressure in a BWR is controlled by the main turbine or main steam bypass valves. Unlike a PWR, where the turbine steam demand is set manually by the operators, in a BWR, the turbine valves will modulate to maintain reactor pressure at a setpoint. Under this control mode, the turbine will automatically follow reactor power changes. When the turbine is offline or trips, the main steam bypass/dump valves will open to direct steam directly to the condenser. These bypass valves will automatically or manually modulate as necessary to maintain reactor pressure and control the reactor's heatup and cooldown rates while steaming is still in progress.

Reactor water level is controlled by the main feedwater system. From about 0.5% power to 100% power, feedwater will automatically control the water level in the reactor. At low power conditions, the feedwater controller acts as a simple PID control by watching reactor water level. At high power conditions, the controller is switched to a "Three-Element" control mode, where the controller looks at the current water level in the reactor, as well as the amount of water going in and the amount of steam leaving the reactor. By using the water injection and steam flow rates, the feed water control system can rapidly anticipate water level deviations and respond to maintain water level within a few inches of set point. If one of the two feedwater pumps fails during operation, the feedwater system will command the recirculation system to rapidly reduce core flow, effectively reducing reactor power from 100% to 50% in a few seconds. At this power level a single feedwater pump can maintain the core water level. If all feedwater is lost, the reactor will scram and the Emergency Core Cooling System is used to restore reactor water level.

Steam Turbines

Steam produced in the reactor core passes through steam separators and dryer plates above the core and then directly to the turbine, which is part of the reactor circuit. Because the water around the core of a reactor is always contaminated with traces of radionuclides, the turbine must be shielded during normal operation, and radiological protection must be provided during maintenance. The increased cost related to operation and maintenance of a BWR tends to balance the savings due to the simpler design and greater thermal efficiency of a BWR when compared with a PWR. Most of the radioactivity in the water is very short-lived (mostly N-16, with a 7-second half-life), so the turbine hall can be entered soon after the reactor is shut down.

BWR steam turbines employ a high-pressure turbine designed to handle saturated steam, and multiple low-pressure turbines. The high-pressure turbine receives steam directly from the reactor. The high-pressure turbine exhaust passes through a steam reheater which superheats the steam to over 400 degrees F for the low-pressure turbines to use. The exhaust of the low-pressure turbines is sent to the main condenser. The steam reheaters take some of the reactor's steam and use it as a heating source to reheat what comes out of the high-pressure turbine exhaust. While the reheaters take steam away from the turbine, the net result is that the reheaters improve the thermodynamic efficiency of the plant.

Reactor Core

A modern BWR fuel assembly comprises 74 to 100 fuel rods, and there are up to approximately 800 assemblies in a reactor core, holding up to approximately 140 short tons of low-enriched uranium. The number of fuel assemblies in a specific reactor is based on considerations of desired reactor power output, reactor core size and reactor power density.

Safety Systems

A modern reactor has many safety systems that are designed with a defence in depth philosophy, which is a design philosophy that is integrated throughout construction and commissioning.

A BWR is similar to a pressurized water reactor (PWR) in that the reactor will continue to produce heat even after the fission reactions have stopped, which could make a core damage incident possible. This heat is produced by the radioactive decay of fission products and materials that have been activated by neutron absorption. BWRs contain multiple safety systems for cooling the core after emergency shut down.

Refueling Systems

The reactor fuel rods are occasionally replaced by removing them from the top of the containment vessel. A typical fuel cycle lasts 18–24 months, with about one third of fuel assemblies being replaced during a refueling outage. The remaining fuel assemblies are shuffled to new core locations to maximize the efficiency and power produced in the next fuel cycle.

Because they are hot both radioactively and thermally, this is done via cranes and under water. For this reason the spent fuel storage pools are above the reactor in typical installations. They are shielded by water several times their height, and stored in rigid arrays in which their geometry is controlled to avoid criticality. In the Fukushima reactor incident this became problematic because water was lost from one or more spent fuel pools and the earthquake could have altered the geometry. The fact that the fuel rods' cladding is a zirconium alloy was also problematic since this element can react with steam at extreme temperatures to produce hydrogen, which can ignite with oxygen in the air. Normally the fuel rods are kept sufficiently cool in the reactor and spent fuel pools that this is not a concern, and the cladding remains intact for the life of the rod.

Evolution

Early Concepts

The BWR concept was developed slightly later than the PWR concept. Development of the BWR started in the early 1950s, and was a collaboration between General Electric (GE) and several US national laboratories.

Research into nuclear power in the US was led by the 3 military services. The Navy, seeing the possibility of turning submarines into full-time underwater vehicles, and ships that could steam around the world without refueling, sent their man in engineering, Captain Hyman Rickover to run their nuclear power program. Rickover decided on the PWR route for the Navy, as the early researchers in the field of nuclear power feared that the direct production of steam within a reactor would cause instability, while they knew that the use of pressurized water would definitively work as a means of heat transfer. This concern led to the US's first research effort in nuclear power being devoted to the PWR, which was highly suited for naval vessels (submarines, especially), as space was at a premium, and PWRs could be made compact and high-power enough to fit in such, in any event.

But other researchers wanted to investigate whether the supposed instability caused by boiling water in a reactor core would really cause instability. During early reactor development, a small group of engineers accidentally increased the reactor power level on an experimental reactor to such an extent that the water quickly boiled, this shut down the reactor, indicating the useful self-moderating property in emergency circumstances. In particular, Samuel Untermyer II, a researcher at Argonne National Laboratory, proposed and oversaw a series of experiments: the BORAX experiments—to see if a *boiling water reactor* would be feasible for use in energy production. He found that it was, after subjecting his reactors to quite strenuous tests, proving the safety principles of the BWR.

Following this series of tests, GE got involved and collaborated with INL to bring this technology to market. Larger-scale tests were conducted through the late 1950s/early/mid-1960s that only partially used directly-generated (primary) nuclear boiler system steam to feed the turbine and incorporated heat exchangers for the generation of secondary steam to drive separate parts of the turbines. The literature does not indicate why this was the case, but it was eliminated on production models of the BWR.

First Series of Production

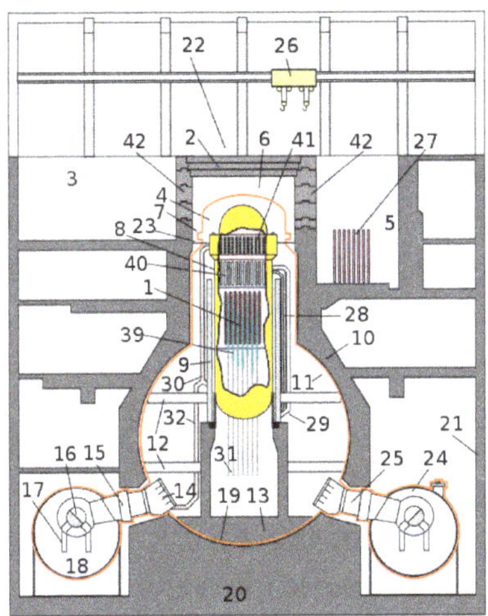

Cross-section sketch of a typical BWR Mark I containment

The first generation of production boiling water reactors saw the incremental development of the unique and distinctive features of the BWR: the torus (used to quench steam in the event of a transient requiring the quenching of steam), as well as the drywell, the elimination of the heat exchanger, the steam dryer, the distinctive general layout of the reactor building, and the standardization of reactor control and safety systems. The first, General Electric (GE), series of production BWRs evolved through 6 iterative design phases, each termed BWR/1 through BWR/6. (BWR/4s, BWR/5s, and BWR/6s are the most common types in service today.) The vast majority of BWRs in service throughout the world belong to one of these design phases.

- 1st generation BWR: BWR/1 with Mark I containment.
- 2nd generation BWRs: BWR/2, BWR/3 and some BWR/4 with Mark I containment. Other BWR/4, and BWR/5 with Mark-II containment.
- 3rd generation BWRs: BWR/6 with Mark-III containment.

Containment variants were constructed using either concrete or steel for the Primary Containment, Drywell and Wetwell in various combinations.

Browns Ferry Unit 1 drywell and wetwell under construction, a BWR/4 using the Mark I containment

Apart from the GE designs there were others by ABB, MITSU, Toshiba and KWU.

Advanced Boiling Water Reactor

A newer design of BWR is known as the Advanced Boiling Water Reactor (ABWR). The ABWR was developed in the late 1980s and early 1990s, and has been further improved to the present day. The ABWR incorporates advanced technologies in the design, including computer control, plant automation, control rod removal, motion, and insertion, in-core pumping, and nuclear safety to deliver improvements over the original series of production BWRs, with a high power output (1350 MWe per reactor), and a significantly lowered probability of core damage. Most significantly, the ABWR was a completely standardized design, that could be made for series production.

The ABWR was approved by the United States Nuclear Regulatory Commission for production as a standardized design in the early 1990s. Subsequently, numerous ABWRs were built in Japan. One development spurred by the success of the ABWR in Japan is that General Electric's nuclear energy division merged with Hitachi Corporation's nuclear energy division, forming GE Hitachi Nuclear Energy, which is now the major worldwide developer of the BWR design.

Simplified Boiling Water Reactor

Parallel to the development of the ABWR, General Electric also developed a different concept, known as the *simplified boiling water reactor* (SBWR). This smaller 600 megawatt electrical reactor was notable for its incorporation—for the first time ever in a light water reactor—of "passive safety" design principles. The concept of passive safety means that the reactor, rather than requiring the intervention of active systems, such as emergency injection pumps, to keep the reactor within safety margins, was instead designed to return to a safe state solely through operation of natural forces if a safety-related contingency developed.

For example, if the reactor got too hot, it would trigger a system that would release soluble neutron absorbers (generally a solution of borated materials, or a solution of borax), or materials that greatly hamper a chain reaction by absorbing neutrons, into the reactor core. The tank containing the soluble neutron absorbers would be located above the reactor, and the absorption solution, once the system was triggered, would flow into the core through force of gravity, and bring the reaction to a near-complete stop. Another example was the Isolation Condenser system, which relied on the principle of hot water/steam rising to bring hot coolant into large heat exchangers located above the reactor in very deep tanks of water, thus accomplishing residual heat removal. Yet another example was the omission of recirculation pumps within the core; these pumps were used in other BWR designs to keep cooling water moving; they were expensive, hard to reach to repair, and could occasionally fail; so as to improve reliability, the ABWR incorporated no less than 10 of these recirculation pumps, so that even if several failed, a sufficient number would remain serviceable so that an unscheduled shutdown would not be necessary, and the pumps could be repaired during the next refueling outage. Instead, the designers of the *simplified boiling water reactor* used thermal analysis to design the reactor core such that natural circulation (cold water falls, hot water rises) would bring water to the center of the core to be boiled.

The ultimate result of the passive safety features of the SBWR would be a reactor that would not require human intervention in the event of a major safety contingency for at least 48 hours following the safety contingency; thence, it would only require periodic refilling of cooling water tanks located completely outside of the reactor, isolated from the cooling system, and designed to remove reactor waste heat through evaporation. The *simplified boiling water reactor* was submitted to the United States Nuclear Regulatory Commission, however, it was withdrawn prior to approval; still, the concept remained intriguing to General Electric's designers, and served as the basis of future developments.

Economic Simplified Boiling Water Reactor

During a period beginning in the late 1990s, GE engineers proposed to combine the features of the *advanced boiling water reactor* design with the distinctive safety features of the *simplified boiling water reactor* design, along with scaling up the resulting design to a larger size of 1,600 MWe (4,500 MWth). This Economic Simplified Boiling Water Reactor (ESBWR) design was submitted to the US Nuclear Regulatory Commission for approval in April 2005, and design certification was granted by the NRC in September 2014.

Reportedly, this design has been advertised as having a core damage probability of only 3×10^{-8} core damage events per reactor-year. That is, there would need to be 3 million ESBWRs operating before one would expect a single core-damaging event during their 100-year lifetimes. Earlier designs of the BWR, the BWR/4, had core damage probabilities as high as 1×10^{-5} core-damage events per reactor-year. This extraordinarily low CDP for the ESBWR far exceeds the other large LWRs on the market.

Advantages and Disadvantages

Advantages

- The reactor vessel and associated components operate at a substantially lower pressure of about 70–75 bars (1,020–1,090 psi) compared to about 155 bars (2,250 psi) in a PWR.
- Pressure vessel is subject to significantly less irradiation compared to a PWR, and so does not become as brittle with age.
- Operates at a lower nuclear fuel temperature.
- Fewer components due to no steam generators and no pressurizer vessel. (Older BWRs have external recirculation loops, but even this piping is eliminated in modern BWRs, such as the ABWR.) This also makes BWRs simpler to operate.
- Lower risk (probability) of a rupture causing loss of coolant compared to a PWR, and lower risk of core damage should such a rupture occur. This is due to fewer pipes, fewer large diameter pipes, fewer welds and no steam generator tubes.
- NRC assessments of limiting fault potentials indicate if such a fault occurred, the average BWR would be less likely to sustain core damage than the average PWR due to the robustness and redundancy of the Emergency Core Cooling System (ECCS).
- Measuring the water level in the pressure vessel is the same for both normal and emergency operations, which results in easy and intuitive assessment of emergency conditions.

- Can operate at lower core power density levels using natural circulation without forced flow.
- A BWR may be designed to operate using only natural circulation so that recirculation pumps are eliminated entirely. (The new ESBWR design uses natural circulation.)
- BWRs do not use boric acid to control fission burn-up to avoid the production of tritium (contamination of the turbines), leading to less possibility of corrosion within the reactor vessel and piping. (Corrosion from boric acid must be carefully monitored in PWRs; it has been demonstrated that reactor vessel head corrosion can occur if the reactor vessel head is not properly maintained. Since BWRs do not utilize boric acid, these contingencies are eliminated.)
- The power control by reduction of the moderator density (vapour bubbles in the water) instead of by addition of neutron absorbers (boric acid in PWR) leads to breeding of U-238 by fast neutrons, producing fissile Pu-239.
- BWRs generally have *N*-2 redundancy on their major safety-related systems, which normally consist of four "trains" of components. This generally means that up to two of the four components of a safety system can fail and the system will still perform if called upon.
- Due to their single major vendor (GE/Hitachi), the current fleet of BWRs have predictable, uniform designs that, while not completely standardized, generally are very similar to one another. The ABWR/ESBWR designs are completely standardized. Lack of standardization remains a problem with PWRs, as, at least in the United States, there are three design families represented among the current PWR fleet (Combustion Engineering, Westinghouse, and Babcock & Wilcox), within these families, there are quite divergent designs. Still, some countries could reach a high level of standardisation with PWRs, like France.
    - Additional families of PWRs are being introduced. For example, Mitsubishi's APWR, Areva's US-EPR, and Westinghouse's AP1000/AP600 will add diversity and complexity to an already diverse crowd, and possibly cause customers seeking stability and predictability to seek other designs, such as the BWR.
- BWRs are overrepresented in imports, when the importing nation does not have a nuclear navy (PWRs are favored by nuclear naval states due to their compact, high-power design used on nuclear-powered vessels; since naval reactors are generally not exported, they cause national skill to be developed in PWR design, construction, and operation). This may be due to the fact that BWRs are ideally suited for peaceful uses like power generation, process/industrial/district heating, and desalinization, due to low cost, simplicity, and safety focus, which come at the expense of larger size and slightly lower thermal efficiency.
    - Sweden is standardized mainly on BWRs.
    - Mexico's two reactors are BWRs.

- o Japan experimented with both PWRs and BWRs, but most builds as of late have been of BWRs, specifically ABWRs.
- o In the CEGB open competition in the early 1960s for a standard design for UK 2nd-generation power reactors, the PWR didn't even make it to the final round, which was a showdown between the BWR (preferred for its easily understood design as well as for being predictable and "boring") and the AGR, a uniquely British design; the indigenous design won, possibly on technical merits, possibly due to the proximity of a general election. In the 1980s the CEGB built a PWR, Sizewell B.

Disadvantages

- BWRs require more complex calculations for managing consumption of nuclear fuel during operation due to "two phase (water and steam) fluid flow" in the upper part of the core. This also requires more instrumentation in the reactor core.
- Larger pressure vessel than for a PWR of similar power, with correspondingly higher cost, in particular for older models that still use a main steam generator and associated piping.
- Contamination of the turbine by short-lived activation products. This means that shielding and access control around the steam turbine are required during normal operations due to the radiation levels arising from the steam entering directly from the reactor core. This is a moderately minor concern, as most of the radiation flux is due to Nitrogen-16 (activation of oxygen in the water), which has a half-life of 7 seconds, allowing the turbine chamber to be entered into within minutes of shutdown.
- Though the present fleet of BWRs is said to be less likely to suffer core damage from the "1 in 100,000 reactor-year" limiting fault than the present fleet of PWRs, (due to increased ECCS robustness and redundancy) there have been concerns raised about the pressure containment ability of the as-built, unmodified Mark I containment – that such may be insufficient to contain pressures generated by a limiting fault combined with complete ECCS failure that results in extremely severe core damage. In this double failure scenario, assumed to be extremely unlikely prior to the Fukushima I nuclear accidents, an unmodified Mark I containment can allow some degree of radioactive release to occur. This is supposed to be mitigated by the modification of the Mark I containment; namely, the addition of an outgas stack system that, if containment pressure exceeds critical setpoints, is supposed to allow the orderly discharge of pressurizing gases after the gases pass through activated carbon filters designed to trap radionuclides.
- Control rods are inserted from below for current BWR designs. There are two available hydraulic power sources that can drive the control rods into the core for a BWR under emergency conditions. There is a dedicated high pressure hydraulic accumulator and also the pressure inside of the reactor pressure vessel available to each control rod. Either the dedicated accumulator (one per rod) or reactor pressure is capable of fully inserting each rod. Most other reactor types use top entry control rods that are held up in the withdrawn position by electromagnets, causing them to fall into the reactor by gravity if power is lost.

Technical and Background Information

Start-up ("Going Critical")

Reactor start up (criticality) is achieved by withdrawing control rods from the core to raise core reactivity to a level where it is evident that the nuclear chain reaction is self-sustaining. This is known as "going critical". Control rod withdrawal is performed slowly, as to carefully monitor core conditions as the reactor approaches criticality. When the reactor is observed to become slightly super-critical, that is, reactor power is increasing on its own, the reactor is declared critical.

Rod motion is performed using rod drive control systems. Newer BWRs such as the ABWR and ESBWR as well as all German and Swedish BWRs use the Fine Motion Control Rod Drive system, which allows multiple rods to be controlled with very smooth motions. This allows a reactor operator to evenly increase the core's reactivity until the reactor is critical. Older BWR designs use a manual control system, which is usually limited to controlling one or four control rods at a time, and only through a series of notched positions with fixed intervals between these positions. Due to the limitations of the manual control system it is possible while starting-up that the core can be placed into a condition where a single control rod can cause a large uneven reactivity change which can potentially challenge the fuel's thermal design margins. As a result, GE developed a set of rules in 1977 called BPWS (Banked Position Withdrawal Sequence) which help minimize the worth of any single control rod and prevent fuel damage in the case of a control rod drop accident. BPWS separates control rods into four groups, A1, A2, B1, and B2. Then, either all of the A control rods or B control rods are pulled full out in a defined sequence to create a "checkboard" pattern. Next the opposing group (B or A) is pulled in a defined sequence to positions 02, then 04, 08, 16, and finally full out (48), until the reactor enters the power operation range where thermal limits are no longer bounding. By following a BPWS compliant start-up sequence, the manual control system can be used to evenly and safely raise the entire core to critical, and prevent any fuel rods from exceeding 280 cal/gm energy release during any postulated event which could potentially damage the fuel.

Thermal Margins

Several calculated/measured quantities are tracked while operating a BWR:

- Maximum Fraction Limiting Critical Power Ratio, or MFLCPR;
- Fraction Limiting Linear Heat Generation Rate, or FLLHGR;
- Average Planar Linear Heat Generation Rate, or APLHGR;
- Pre-Conditioning Interim Operating Management Recommendation, or PCIOMR;

MFLCPR, FLLHGR, and APLHGR must be kept less than 1.0 during normal operation; administrative controls are in place to assure some margin of error and margin of safety to these licensed limits. Typical computer simulations divide the reactor core into 24–25 axial planes; relevant quantities (margins, burnup, power, void history) are tracked for each "node" in the reactor core (764 fuel assemblies x 25 nodes/assembly = 19100 nodal calculations/quantity).

Maximum Fraction Limiting Critical Power Ratio (MFLCPR)

Specifically, MFLCPR represents how close the leading fuel bundle is to "dry-out" (or "departure from nucleate boiling" for a PWR). Transition boiling is the unstable transient region where nucleate boiling tends toward film boiling. A water drop dancing on a hot frying pan is an example of film boiling. During film boiling a volume of insulating vapor separates the heated surface from the cooling fluid; this causes the temperature of the heated surface to increase drastically to once again reach equilibrium heat transfer with the cooling fluid. In other words, steam semi-insulates the heated surface and surface temperature rises to allow heat to get to the cooling fluid (through convection and radiative heat transfer).

MFLCPR is monitored with an empirical correlation that is formulated by vendors of BWR fuel (GE, Westinghouse, AREVA-NP). The vendors have test rigs where they simulate nuclear heat with resistive heating and determine experimentally what conditions of coolant flow, fuel assembly power, and reactor pressure will be in/out of the transition boiling region for a particular fuel design. In essence, the vendors make a model of the fuel assembly but power it with resistive heaters. These mock fuel assemblies are put into a test stand where data points are taken at specific powers, flows, pressures. It is obvious that nuclear fuel could be damaged by film boiling; this would cause the fuel cladding to overheat and fail. Experimental data is conservatively applied to BWR fuel to ensure that the transition to film boiling does not occur during normal or transient operation. Typical SLMCPR/MCPRSL (Safety Limit MCPR) licensing limit for a BWR core is substantiated by a calculation that proves that 99.9% of fuel rods in a BWR core will not enter the transition to film boiling during normal operation or anticipated operational occurrences. Since the BWR is boiling water, and steam does not transfer heat as well as liquid water, MFLCPR typically occurs at the top of a fuel assembly, where steam volume is the highest.

Fraction Limiting Linear Heat Generation Rate (FLLHGR)

FLLHGR (FDLRX, MFLPD) is a limit on fuel rod power in the reactor core. For new fuel, this limit is typically around 13 kW/ft (43 kW/m) of fuel rod. This limit ensures that the centerline temperature of the fuel pellets in the rods will not exceed the melting point of the fuel material (uranium/gadolinium oxides) in the event of the worst possible plant transient/scram anticipated to occur. To illustrate the response of LHGR in transient imagine the rapid closure of the valves that admit steam to the turbines at full power. This causes the immediate cessation of steam flow and an immediate rise in BWR pressure. This rise in pressure effectively subcools the reactor coolant instantaneously; the voids (vapor) collapse into solid water. When the voids collapse in the reactor, the fission reaction is encouraged (more thermal neutrons); power increases drastically (120%) until it is terminated by the automatic insertion of the control rods. So, when the reactor is isolated from the turbine rapidly, pressure in the vessel rises rapidly, which collapses the water vapor, which causes a power excursion which is terminated by the Reactor Protection System. If a fuel pin was operating at 13.0 kW/ft prior to the transient, the void collapse would cause its power to rise. The FLLHGR limit is in place to ensure that the highest powered fuel rod will not melt if its power was rapidly increased following a pressurization transient. Abiding by the LHGR limit precludes melting of fuel in a pressurization transient.

Average Planar Linear Heat Generation Rate (APLHGR)

APLHGR, being an average of the Linear Heat Generation Rate (LHGR), a measure of the decay heat present in the fuel bundles, is a margin of safety associated with the potential for fuel failure to occur during a LBLOCA (large-break loss-of-coolant accident – a massive pipe rupture leading to catastrophic loss of coolant pressure within the reactor, considered the most threatening "design basis accident" in probabilistic risk assessment and nuclear safety), which is anticipated to lead to the temporary exposure of the core; this core drying-out event is termed core "uncovery", for the core loses its heat-removing cover of coolant, in the case of a BWR, light water. If the core is uncovered for too long, fuel failure can occur; for the purpose of design, fuel failure is assumed to occur when the temperature of the uncovered fuel reaches a critical temperature (1100 °C, 2200 °F). BWR designs incorporate failsafe protection systems to rapidly cool and make safe the uncovered fuel prior to it reaching this temperature; these failsafe systems are known as the Emergency Core Cooling System. The ECCS is designed to rapidly flood the reactor pressure vessel, spray water on the core itself, and sufficiently cool the reactor fuel in this event. However, like any system, the ECCS has limits, in this case, to its cooling capacity, and there is a possibility that fuel could be designed that produces so much decay heat that the ECCS would be overwhelmed and could not cool it down successfully.

So as to prevent this from happening, it is required that the decay heat stored in the fuel assemblies at any one time does not overwhelm the ECCS. As such, the measure of decay heat generation known as LHGR was developed by GE's engineers, and from this measure, APLHGR is derived. APLHGR is monitored to ensure that the reactor is not operated at an average power level that would defeat the primary containment systems. When a refueled core is licensed to operate, the fuel vendor/licensee simulate events with computer models. Their approach is to simulate worst case events when the reactor is in its most vulnerable state.

APLHGR is commonly pronounced as "Apple Hugger" in the industry.

Pre-conditioning Interim Operating Management Recommendation (PCIOMR)

PCIOMR is a set of rules and limits to prevent cladding damage due to pellet-clad interaction. During the first nuclear heatup, nuclear fuel pellets can crack. The jagged edges of the pellet can rub and interact with the inner cladding wall. During power increases in the fuel pellet, the ceramic fuel material expands faster than the fuel cladding, and the jagged edges of the fuel pellet begin to press into the cladding, potentially causing a perforation. To prevent this from occurring, two corrective actions were taken. The first is the inclusion of a thin barrier layer against the inner walls of the fuel cladding which are resistant to perforation due to pellet-clad interactions, and the second is a set of rules created under PCIOMR.

The PCIOMR rules require initial "conditioning" of new fuel. This means, for the first nuclear heatup of each fuel element, that local bundle power must be ramped very slowly to prevent cracking of the fuel pellets and limit the differences in the rates of thermal expansion of the fuel. PCIOMR rules also limit the maximum local power change (in kW/ft*hr), prevent pulling control rods below the tips of adjacent control rods, and require control rod sequences to be analyzed against core modelling software to prevent pellet-clad interactions. PCIOMR analysis look at local power peaks and xenon transients which could be caused by control rod position changes or rapid power changes to ensure that local power rates never exceed maximum ratings.

List of BWRs

For a list of operational and decommissioned BWRs.

Experimental and Other Types

Experimental and other non-commercial BWRs include:

- BORAX experiments
- EBWR (Experimental Boiling Water Reactor)
- SL-1 (destroyed during accident in 1961)

Next-generation Designs

- Advanced Boiling Water Reactor (ABWR)
- Economic Simplified Boiling Water Reactor (ESBWR)
- Areva Kerena (Based on Siemens SWR 1000, Siemens sold its nuclear business to Areva)
- Toshiba ABWR (Not related to GE-Hitachi ABWR, Based on Asea (now part of ABB) BWR 90+ design, ABB exited the nuclear business and the design is now owned by Toshiba via a series of mergers and divestment of nuclear business. Asea→ABB→Westinghouse→Toshiba)

Pressurized Heavy-water Reactor

A pressurized heavy-water reactor (PHWR) is a nuclear reactor, commonly using unenriched natural uranium as its fuel, that uses heavy water (deuterium oxide D_2O) as its coolant and neutron moderator. The heavy water coolant is kept under pressure, allowing it to be heated to higher temperatures without boiling, much as in a pressurized water reactor. While heavy water is significantly more expensive than ordinary light water, it creates greatly enhanced neutron economy, allowing the reactor to operate without fuel-enrichment facilities (offsetting the additional expense of the heavy water) and enhancing the ability of the reactor to make use of alternate fuel cycles.

Purpose of using Heavy Water

The key to maintaining a nuclear reaction within a nuclear reactor is to use the neutrons released during fission to stimulate fission in other nuclei. With careful control over the geometry and reaction rates, this can lead to a self-sustaining chain reaction, a state known as "criticality".

Natural uranium consists of a mixture of various isotopes, primarily ^{238}U and a much smaller amount (about 0.72% by weight) of ^{235}U. ^{238}U can only be fissioned by neutrons that are relatively energetic, about 1 MeV or above. No amount of ^{238}U can be made "critical" since it will tend to parasitically absorb more neutrons than it releases by the fission process. ^{235}U, on the other hand, can support a self-sustained chain reaction, but due to the low natural abundance of ^{235}U, natural uranium cannot achieve criticality by itself.

The "trick" to making a working reactor fuelled by natural or low enriched Uranium is to slow enough of the neutrons to the point where their probability of causing nuclear fission in ^{235}U increases to a level that permits a sustained chain reaction in the uranium as a whole. This requires the use of a neutron moderator, which absorbs some of the neutrons' kinetic energy, slowing them down to an energy comparable to the thermal energy of the moderator nuclei themselves (leading to the terminology of "thermal neutrons" and "thermal reactors"). During this slowing-down process it is beneficial to physically separate the neutrons from the uranium, since ^{238}U nuclei have an enormous parasitic affinity for neutrons in this intermediate energy range (a reaction known as "resonance" absorption). This is a fundamental reason for designing reactors with discrete solid fuel separated by moderator, rather than employing a more homogeneous mixture of the two materials.

Water makes an excellent moderator; the hydrogen atoms in the water molecules are very close in mass to a single neutron, and the collisions thus have a very efficient momentum transfer, similar conceptually to the collision of two billiard balls. However, despite being a good moderator, water is relatively effective at absorbing neutrons. Using water as a moderator will absorb so many neutrons that there will be too few left to react with the small amount of ^{235}U in the fuel, thus precluding criticality in natural uranium. Instead, in order to fuel a light-water reactor, first the amount of ^{235}U in the uranium must be increased, producing enriched uranium, which generally contains between 3% and 5% ^{235}U by weight (the waste from this process is known as depleted uranium, consisting primarily of ^{238}U). In this enriched form there *is* enough ^{235}U to react with the water-moderated neutrons to maintain criticality.

One complication of this approach is the need for uranium enrichment facilities, which are generally expensive to build and operate. They also present a nuclear proliferation concern; the same systems used to enrich the ^{235}U can also be used to produce much more "pure" weapons-grade material (90% or more ^{235}U), suitable for producing a nuclear bomb. This is not a trivial exercise by any means, but feasible enough that enrichment facilities present a significant nuclear proliferation risk.

An alternative solution to the problem is to use a moderator that does *not* absorb neutrons as readily as water. In this case potentially all of the neutrons being released can be moderated and used in reactions with the ^{235}U, in which case there *is* enough ^{235}U in natural uranium to sustain criticality. One such moderator is heavy water, or deuterium-oxide. Although it reacts dynamically with the neutrons in a fashion similar to light water (albeit with less energy transfer on average, given that heavy hydrogen, or deuterium, is about twice the mass of hydrogen), it already has the extra neutron that light water would normally tend to absorb.

Advantages and Disadvantages

The use of heavy water as the moderator is the key to the PHWR (pressurized heavy water reactor) system, enabling the use of natural uranium as the fuel (in the form of ceramic UO_2), which means that it can be operated without expensive uranium enrichment facilities. The mechanical arrangement of the PHWR, which places most of the moderator at lower temperatures, is particularly efficient because the resulting thermal neutrons are "more thermal" than in traditional designs, where the moderator normally is much hotter. These features mean that a PHWR can use natural uranium and other fuels, and does so more efficiently than light water reactors (LWRs).

Pressurised heavy-water reactors do have some drawbacks. Heavy water generally costs hundreds of dollars per kilogram, though this is a trade-off against reduced fuel costs. The reduced energy content of natural uranium as compared to enriched uranium necessitates more frequent replacement of fuel; this is normally accomplished by use of an on-power refuelling system. The increased rate of fuel movement through the reactor also results in higher volumes of spent fuel than in LWRs employing enriched uranium. However, since unenriched uranium fuel accumulates a lower density of fission products than enriched uranium fuel, it generates less heat, allowing more compact storage.

Nuclear Proliferation

Opponents of heavy-water reactors suggest that such reactors pose a much greater risk of nuclear proliferation than comparable light water reactors. These concerns stem from the fact that during normal reactor operation, uranium-238 in the natural uranium fuel of a heavy-water reactor is converted into plutonium-239 (via neutron capture followed by two β^- decays). Plutonium-239 is a fissile material suitable for use in nuclear weapons. As a result, if the fuel of a heavy-water reactor is changed frequently, significant amounts of weapons-grade plutonium can be chemically extracted from the irradiated natural uranium fuel by nuclear reprocessing. In this way, the materials necessary to construct a nuclear weapon can be obtained without any uranium enrichment.

In addition, the use of heavy water as a moderator results in the production of small amounts of tritium when the deuterium nuclei in the heavy water absorb neutrons, a very inefficient reaction. Tritium is essential for the production of boosted fission weapons, which in turn enable the easier production of thermonuclear weapons, including neutron bombs. It is unclear whether it is possible to use this method to produce tritium on a practical scale.

The proliferation risk of heavy-water reactors was demonstrated when India produced the plutonium for Operation Smiling Buddha, its first nuclear weapon test, by extraction from the spent fuel of a heavy-water research reactor known as the CIRUS reactor.

Gas-cooled Reactor

A schematic diagram of a gas-cooled reactor is shown in Figure. This is a type of nuclear reactor that uses a gas as the coolant. Mostly CO_2 is used. Graphite blocks are used as moderator, within which channels are made for housing fuel rods. Control rods are inserted into the graphite blocks. Channels are established between the graphite blocks for the flow of coolant. Natural uranium is used as the fuel while cladding is made of a magnesium alloy called magnox.This reactor derives its name from the alloy used for cladding, ‘magnox’. The coolant gas (CO_2) is supplied by a gas circulator and enters the core from bottom. Gas flows through the coolant channels between the graphite blocks. As the gas moves up through the core, it gets heated up and leaves the top of the core at high temperature.

This high temperature gas exchanges heat with water in a heat exchanger, resulting in the production of steam, which runs the turbine. The spent steam is condensed and returned back to the heat exchanger, while the gas returns to the reactor. The heat exchanger is located outside the pressure vessel and the containment. Magnox reactor finds extensive use in United Kingdom.

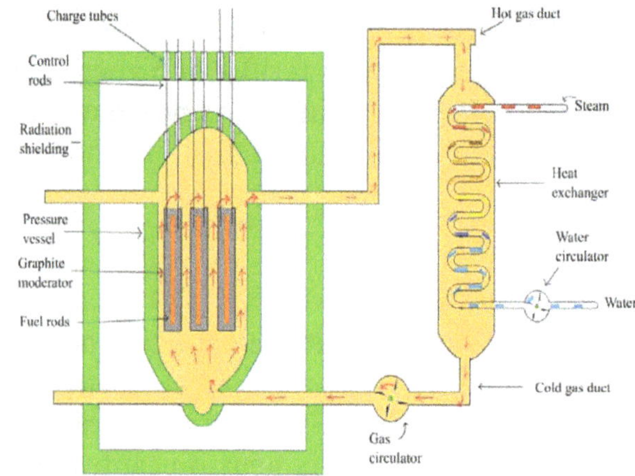

Schematic diagram of gas-cooled reactor (Redrawn from Ref.)

Advanced Gas-cooled Reactor (AGCR)

A major difference between the GCR and AGCR is the location of heat exchangers (2 in number) within the reactor pressure vessel and the containment in AGCR. The other differences are the use of stainless steel fuel cladding and the use of enriched fuel (2.5 – 3.5 % U-235). Higher energy per unit mass of the fuel is obtained owing to higher fuel enrichment. Due to the integration of heat exchanger inside the reactor vessel and in a pool of hot gas, higher thermal efficiencies are achieved compared to that in a GCR.

An Advanced Gas-cooled Reactor (AGR) is a specific type of nuclear reactor. These are the second generation of British gas-cooled reactors, using graphite as the neutron moderator and carbon dioxide as coolant. The AGR was developed from the Magnox reactor, and operates at a higher gas temperature for improved thermal efficiency, but requires stainless steel fuel cladding to withstand the higher temperature. Because the stainless steel fuel cladding has a higher neutron capture cross section than Magnox fuel cans, enriched uranium fuel is needed, with the benefit of higher "burn ups" of 18,000 MW_t-days per tonne of fuel, requiring less frequent refuelling. The first prototype AGR became operational in 1962 but the first commercial AGR did not come online until 1976.

AGR power station at Torness

All existing AGR power stations are configured with two reactors in a single building. Each reactor has a design thermal power output of 1,500 MW_t driving a 660 MW_e turbine-alternator set. The various AGR stations produce outputs in the range 555 MWe to 670 MWe though some run at lower than design output due to operational restrictions.

AGR Design

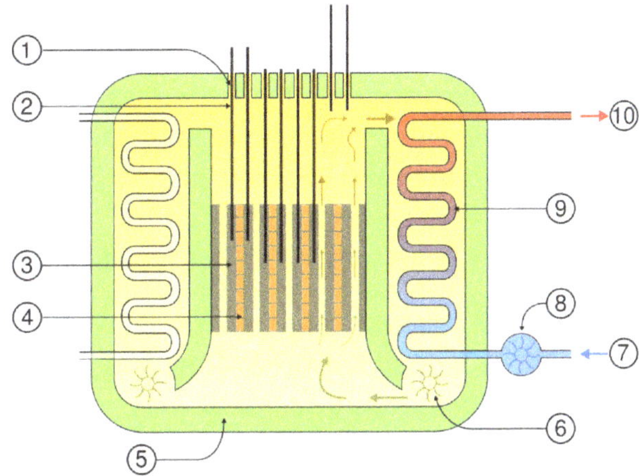

Schematic diagram of the Advanced Gas-cooled Reactor. Note that the heat exchanger is contained within the steel-reinforced concrete combined pressure vessel and radiation shield. 1. Charge tubes 2. Control rods 3. Graphite moderator 4. Fuel assemblies 5. Concrete pressure vessel and radiation shielding 6. Gas circulator 7. Water 8. Water circulator 9. Heat exchanger 10. Steam

The two power stations with four AGRs at Heysham

The design of the AGR was such that the final steam conditions at the boiler stop valve were identical to that of conventional coal-fired power stations, thus the same design of turbo-generator plant could be used. The mean temperature of the hot coolant leaving the reactor core was designed to be 648 °C. In order to obtain these high temperatures, yet ensure useful graphite core life (graphite oxidises readily in CO_2 at high temperature) a re-entrant flow of coolant at the lower boiler outlet temperature of 278 °C is utilised to cool the graphite, ensuring that the graphite core temperatures do not vary too much from those seen in a Magnox station. The superheater outlet temperature and pressure were designed to be 2,485 psi (170 bar) and 543 °C.

The fuel is uranium dioxide pellets, enriched to 2.5-3.5%, in stainless steel tubes. The original design concept of the AGR was to use a beryllium based cladding. When this proved unsuitable due to brittle fracture, the enrichment level of the fuel was raised to allow for the higher neutron capture losses of stainless steel cladding. This significantly increased the cost of the power produced by an AGR. The carbon dioxide coolant circulates through the core, reaching 640 °C (1,184 °F) and a pressure of around 40 bar (580 psi), and then passes through boiler (steam generator) assemblies outside the core but still within the steel-lined, reinforced concrete pressure vessel. Control rods penetrate the graphite moderator and a secondary system involves injecting nitrogen into the coolant to hold the reactor temperature down. A tertiary shutdown system which operates by injecting boron balls into the reactor is included in case the reactor has to be depressurized with insufficient control rods lowered. This would mean that nitrogen pressure can not be maintained.

The AGR was designed to have a high thermal efficiency (electricity generated/heat generated ratio) of about 41%, which is better than modern pressurized water reactors which have a typical thermal efficiency of 34%. This is due to the higher coolant outlet temperature of about 640 °C (1,184 °F) practical with gas cooling, compared to about 325 °C (617 °F) for PWRs. However the reactor core has to be larger for the same power output, and the fuel burnup ratio at discharge is lower so the fuel is used less efficiently, countering the thermal efficiency advantage.

Like the Magnox, CANDU and RBMK reactors, and in contrast to the light water reactors, AGRs are designed to be refuelled without being shut down first. This on-load refuelling was an important part of the economic case for choosing the AGR over other reactor types, and in 1965 allowed the Central Electricity Generating Board (CEGB) and the government to claim that the AGR would produce electricity cheaper than the best coal-fired power stations. However fuel assembly vibration problems arose during on-load refuelling at full power, so in 1988 full power refuelling was suspended until the mid-1990s, when further trials led to a fuel rod becoming stuck in a reactor core. Only refuelling at part load or when shut down is now undertaken at AGRs.

The AGR was intended to be a superior British alternative to American light water reactor designs. It was promoted as a development of the operationally (if not economically) successful Magnox design, and was chosen from a multitude of competing British alternatives - the helium cooled High Temperature Reactor (HTR), the Steam Generating Heavy Water Reactor (SGHWR) and the Fast Breeder Reactor (FBR) - as well as the American light water pressurised and boiling water reactors (PWR and BWR) and Canadian CANDU designs. The CEGB conducted a detailed economic appraisal of the competing designs and concluded that the AGR proposed for Dungeness B would generate the cheapest electricity, cheaper than any of the rival designs and the best coal-fired stations.

History

There were great hopes for the AGR design. An ambitious construction programme of five twin reactor stations, Dungeness B, Hinkley Point B, Hunterston B, Hartlepool and Heysham was quickly rolled out, and export orders were eagerly anticipated. However, the AGR design proved to be over complex and difficult to construct on site. Notoriously bad labour relations at the time added to the problems. The lead station, Dungeness B was ordered in 1965 with a target completion date of 1970. After problems with nearly every aspect of the reactor design it finally began generating electricity in 1983, 13 years late. The following reactor designs at Hinkley Point and Hunterston significantly improved on the original design and indeed were commissioned ahead of Dungeness. The next AGR design at Heysham 1 and Hartlepool sought to reduce overall cost of design by reducing the footprint of the station and the number of ancillary systems. The final two AGRs at Torness and Heysham 2 returned to a modified Hinkley design and have proved to be the most successful performers of the fleet. Former Treasury Economic Advisor, David Henderson, described the AGR programme as one of the two most costly British government-sponsored project errors, alongside Concorde.

When the government started on privatising the electricity generation industry in the 1980s, a cost analysis for potential investors revealed that true operating costs had been obscured for many years. Decommissioning costs especially had been significantly underestimated. These uncertainties caused nuclear power to be omitted from the privatisation at that time.

The small-scale prototype AGR at Sellafield (Windscale) has been decommissioned as of 2010 – the core and pressure vessel decommissioned leaving only the building "Golf Ball" visible. This project was also a study of what is required to decommission a nuclear reactor safely.

In October 2016 it was announced that super-articulated control rods would be installed at Hunterston B and Hinkley Point B because of concerns about the stability of the reactors' graphite cores. The Office for Nuclear Regulation (ONR) had raised concerns over the number of fractures in keyways that lock together the graphite bricks in the core. An unusual event, such as an earthquake, might destabilise the graphite so that ordinary control rods that shut the reactor down could not be inserted. Super-articulated control rods should be insertable even into a destabilised core.

Current AGR Reactors

As of 22 May 2010, there are seven nuclear generating stations each with two operating AGRs in the United Kingdom, owned and operated by EDF Energy:

AGR Power Station	Net MWe	Construction started	Connected to grid	Commercial operation	Accounting closure date
Dungeness B	1110	1965	1983	1985	2028
Hartlepool	1210	1968	1983	1989	2024
Heysham 1	1150	1970	1983	1989	2024
Heysham 2	1250	1980	1988	1989	2030
Hinkley Point B	1220	1967	1976	1976	2023
Hunterston B	1190	1967	1976	1976	2023
Torness	1250	1980	1988	1988	2030

In 2005 British Energy announced a 10-year life extension at Dungeness B, that will see the station continue operating until 2018, and in 2007 announced a 5-year life extension of Hinkley Point B and Hunterston B until 2016. Life extensions at other AGRs will be considered at least three years before their scheduled closure dates.

From 2006 Hinkley Point B and Hunterston B have been restricted to about 70% of normal MWe output because of boiler-related problems requiring that they operate at reduced boiler temperatures. In 2013 these two stations' power increased to about 80% of normal output following some plant modifications.

In 2006 AGRs made the news when documents were obtained under the Freedom of Information Act 2000 by The Guardian who claimed that British Energy were unaware of the extent of the cracking of graphite bricks in the cores of their reactors. It was also claimed that British Energy did not know why the cracking had occurred and that they were unable to monitor the cores without first shutting down the reactors. British Energy later issued a statement confirming that cracking of graphite bricks is a known symptom of extensive neutron bombardment and that they were working on a solution to the monitoring problem. Also, they stated that the reactors were examined every three years as part of "statutory outages".

On 17 December 2010, EDF Energy announced a 5-year life extension for both Heysham 1 and Hartlepool to enable further generation until 2019.

In February 2012 EDF announced it expects 7 year life extensions on average across all AGRs, including the recently life-extended Heysham 1 and Hartlepool. These life extensions are subject to detailed review and approval, and are not included in the table above.

On 4 December 2012 EDF announced that Hinkley Point B and Hunterston B had been given 7 year life extensions, from 2016 to 2023.

On 5 November 2013 EDF announced that Hartlepool had been given a 5-year life extension, from 2019 to 2024.

In 2013 a defect was found by a regular inspection in one of the eight pod boilers of Heysham reactor A1. The reactor resumed operation at a lower output level with this pod boiler disabled, until June 2014 when more detailed inspections confirmed a crack in the boiler spine. As a precaution Heysham A2 and the sister Hartlepool station were also closed down for an eight weeks' inspection.

In October 2014 a new kind of crack in the graphite moderator bricks was found at the Hunterston B reactor. This keyway root crack has been previously theorized but not observed. The existence of this type of crack does not immediately affect the safety of a reactor – however if the number of cracks exceed a threshold the reactor would be decommissioned, as the cracks cannot be repaired.

In January 2015 Dungeness B was given a ten-year life extension, with an upgrade to control room computer systems and improved flood defences, taking the accounting closure date to 2028.

In February 2016, EDF extended the life of four of its eight nuclear power plants in the UK. Heysham 1 and Hartlepool will have their life extended by five years until 2024, while Heysham 2 and Torness will see their closure dates pushed back by seven years to 2030.

High-Temperature Gas Reactor (HTGR)

High-temperature gas reactor uses Helium gas as the coolant, owing to its chemical inertness and better thermodynamic properties. Fine particles of UO_2 (~ 0.5 mm) are coated with layers of porous carbon, pyrolytic carbon, silicon carbide and pyrolytic carbon again. Coating with silicon carbide provides protection against melt down up to 1600 °C. The diameter of coated particles is about 0.92 mm. This type of fuel is called TRISO (Tristructural-isotropic) fuel.

If these coated particles are embedded in spherical graphite matrices of 60 mm diameter to form pebbles, and used in the reactor with the voids created by the packing of pebbles making the path for coolant (helium gas) flow, the reactor is called Pebble Bed Reactor (PBR). The moderator (graphite) is integrated with the fuel such that the separation of moderator from fuel never occurs. There are no control rods in these reactors for power control. Increase in temperature of the core causes the fission rate to decrease by a phenomenon called 'Doppler broadening'. This ensures that the heat generation does not exceed beyond design limits.

The heat extracted by helium may be directly used to operate a high temperature gas turbine or can be used to generate steam by extracting heat from helium in a heat exchanger/steam generator.If the fuel particles are compacted and placed in a graphite block (as in a Gas-Cooled Reactor), the reactor is called prismatic-block-gas-cooled reactor. An advantage of PBR over prismatic block reactor is the facility for continuous replacement of fuel in PBR. South Africa is the pioneer of PBR technology.

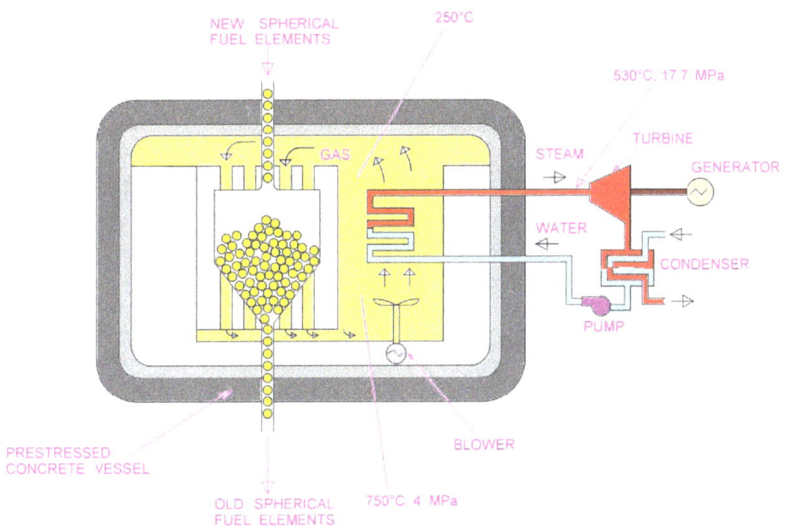

Schematic diagram of pebble bed reactor (Modified from Ref.)

There were two commercial HTGR plants: (i) 330 MWe plant at Fort St. Vrain in USA and (ii) 300 MWe (THTR-300) plant at Germany. These were in operation from early 1970 till 1990and early 1980 till 1990 respectively. The two prototype HTGR in operation, as on 2011, were HTTR (Japan) and HTR-10 (China) with the capacities of 30 MWt and 10 MWt respectively.

Advanced Heavy Water Reactor (AHWR)

This reactor is an Indian version of next generation Heavy Water Reactor. Boiling water is the coolant while heavy water is the moderator. The proposed reactor design is of vertical, pressure tube type rated at 920 MWTh and 300 MWe. This reactor is designed to utilize U-233 as the fissile isotope, the initial loading obtained from breeding in fast breeder reactor by the nuclear transmutation of Th-232 to U-233. At the centre of a typical fuel cluster is a displacer rod containing dysprosia in a zirconia matrix along with a water tube for the injection of water from Emergency Core Cooling System (ECCS) on to fuel pins directly. Surrounding the central rod are two rows of (Th-232 & U-233) oxide fuel pins (30 in number, 11.2 mm diameter) and a row of (Th-232 and Pu- 239) oxide fuel pins (24 in number, 11.2 mm diameter). This configuration ensures a slightly negative reactivity coefficient similar to that of BWR. The pressure tubes are 120 mm in diameter. The number of coolant channels is ~ 452. The pressure in the reactor is about 70 bar, with the calendria being 8000 mm in diameter and 5000 mm in length.

By using a mixture of U-233 and Th-232 in the core, simultaneous breeding (Th-232 to U-233) and burning of U-233 can be achieved with thermal neutrons. The capture cross section of Th-232 for thermal neutrons is high to facilitate transmutation and breeding. Hence this proposed reactor is the first-of-its-kind thermal breeder. This reactor facilitates utilization of Thorium that is found in abundance in India.

Heat released during fission is removed by the boiling of light water under natural circulation, making it as an inherently safe passive feature. This circumvents the need for coolant pumps and hence brings about a reduction in cost of primary coolant system. Use of a large water tank on top of the primary containment called Gravity Driven Water Pool (GDWP) is a unique feature of this design that adds to the safety features of the reactor.

AHWR, being a reactor cooled by natural circulation, power density is lower. Hence lower level of neutron flux is required, which serves to increase fertile to fissile fuel conversion.

The important safety features of AHWR are:

(i) Negative void coefficient of reactivity, implying lower heat generation with increase in void fraction (volume fraction occupied by steam)

(ii) Passive safety systems that do not require operator intervention

(iii) Presence of a large 6000 m^3 of water in the form of gravity driven water pool

(iv) Heat removal from core by natural convection during normal operation and shut down.

Fast Reactors

Fast reactors are the ones that utilize fast neutrons for fission. The reactor is devoid of any light nuclei which act as moderators. The use of fast neutrons necessitates the use of Pu-239 as the main fissile isotopes at high enrichment levels. The reactor core is hexagonal, with fuel pins consisting of mixed oxide (PuO_2-UO_2). The power cycle is three-coolant, steam cycle. Sodium is used as coolant in two cycles, while water is used as coolant in third cycle. Use of liquid sodium as coolant permits reactor operation at higher temperatures and subsequently steam generation at higher temperatures. Primary sodium is the sodium in contact with the core and hence is radioactive, while secondary sodium is the one in thermal contact with primary sodium on one side of the loop. On the other side of the loop, secondary sodium (non-radioactive) is in thermal contact with water for steam generation. Two sodium loops are used to prevent even accidental contact of radioactive sodium with water. Steam-water cycle is similar to that used in PWR or PHWR.

References

- Boiling Water Reactor Simulator with Passive Safety Systems - IAEA (PDF (11 MB)), IAEA, October 2009, p. 14, retrieved 8 June 2012
- Murray, P. (1981). "Developments in oxide fuels at Harwell". Journal of Nuclear Materials. 100 (1–3): 67–71. Bibcode:1981JNuM..100...67M. doi:10.1016/0022-3115(81)90521-3
- Susan Dingman; Jeff LaChance; Allen Canip; Mary Drouin. "Core damage frequency perspectives for BWR 3/4 and Westinghouse 4-loop plants based on IPE results". Osti.gov. Retrieved 2013-08-02

- Shultis, J. Kenneth; Richard E. Faw (2002). Fundamentals of Nuclear Science and Engineering. Marcel Dekker. ISBN 0-8247-0834-2
- S H Wearne, R H Bird (December 2016). "UK Experience of Consortia Engineering for Nuclear Power Stations" (PDF). Dalton Nuclear Institute, University of Manchester. Retrieved 25 March 2017
- "Life extension of Hinkley Point B and Hunterston B power stations". British Energy. 11 December 2007. Retrieved 2008-06-19

5

An Overview of Fast Reactors

Nuclear fast reactor is a type of nuclear reactor. Instead of using neutron moderator, nuclear fast reactors use fast neutrons. Water too, is not used as a coolant in these types of reactors. In order to completely understand nuclear science and technology, it is necessary to understand the processes related to it. The following chapter elucidates the varied processes and mechanisms associated with this area of study

Fast-neutron Reactor

Shevchenko BN350 nuclear fast reactor and desalination plant situated on the shore of the Caspian Sea. The plant generated 135 MW_e and provided steam for an associated desalination plant. View of the interior of the reactor hall.

A fast neutron reactor or simply a fast reactor is a category of nuclear reactor in which the fission chain reaction is sustained by fast neutrons. Such a reactor needs no neutron moderator, but must use fuel that is relatively rich in fissile material when compared to that required for a thermal reactor.

Basic Fission Concepts

In order to sustain a fission chain reaction, the neutrons released in fission events have to react with other atoms in the fuel. The chance of this occurring depends on the energy of the neutron; most atoms will only undergo induced fission with high energy neutrons, although a smaller number prefer much lower energies.

Natural uranium consists mostly of three isotopes, U-238, U-235, and trace quantities of U-234, a decay product of U-238. U-238 accounts for roughly 99.3% of natural uranium and undergoes fission only by neutrons with energies of 5 MeV or greater, the so-called fast neutrons. About 0.7%

of natural uranium is U-235, which undergoes fission by neutrons of any energy, but particularly by lower energy neutrons. When either of these isotopes undergoes fission they release neutrons around 1 to 2 MeV, too low to cause fission in U-238, and too high to do so easily in U-235.

The common solution to this problem is to slow the neutron from these fast speeds using a neutron moderator, any substance which interacts with the neutrons and slows their speed. The most common moderator is normal water, which slows the neutrons through elastic scattering until the neutrons reach thermal equilibrium with the water. The key to reactor design is to carefully lay out the fuel and water so the neutrons have time to slow enough to become highly reactive with the U-235, but not so far as to allow them easy pathways to escape the reactor core entirely.

Although U-238 will not undergo fission by the neutrons released in fission, thermal neutrons can be captured by the nucleus to transmute the atom into Pu-239. Pu-239 has a neutron cross section very similar to that of U-235, and most of the atoms created this way will undergo fission from the thermal neutrons. In most reactors this accounts for as much as ⅓ of the energy being generated. Not all of the Pu-239 is burned up during normal operation, and the leftover, along with leftover U-238, can be separated out to be used in new fuel during nuclear reprocessing.

Water is a common moderator for practical reasons, but has its disadvantages. From a nuclear standpoint, the primary problem is that water can absorb a neutron and remove it from the reaction. It does this just enough that the amount of U-235 in natural ore is too low to sustain the chain reaction; the neutrons lost through absorption in the water and U-238, along with those lost to the environment, results in too few left in the fuel. The most common solution to this problem is to slightly concentrate the amount of U-235 in the fuel to produce enriched uranium, with the leftover U-238 known as depleted uranium. Other designs use different moderators, like heavy water, that are much less likely to absorb neutrons, allowing them to run on unenriched fuel. In either case, the reactor's neutron economy is based on thermal neutrons.

Fast Fission, Breeders

Although U-235 and Pu-239 are less sensitive to higher energy neutrons, they still remain somewhat reactive well into the MeV area. This means that if you enrich the fuel you will eventually reach a threshold where there are enough fissile atoms in the fuel that a chain reaction can be maintained even with fast neutrons.

The primary advantage is that by removing the moderator, the size of the reactor can be greatly reduced, and to some extent the complexity. This is commonly used for shipboard and submarine reactor systems, where size and weight are major concerns. The downside to the fast reaction is that fuel enrichment is an expensive process, so this is generally not suitable for electrical generation or other roles where cost is more important than size.

There is another advantage to the fast reaction that has led to considerable development for civilian use. Fast reactors lack a moderator, and thus lack one of the systems that remove neutrons from the system. Those running on Pu-239 further increase the number of neutrons, because its most common fission cycle gives off three neutrons rather than the mix of two and three neutrons released from U-235. By surrounding the reactor core with a moderator and then a *blanket* of U-238, those neutrons can be captured and used to breed more Pu-239. This is the same reaction that occurs internally in conventional designs, but in this case the blanket does not have to sustain a reaction and thus can be made of natural uranium or even depleted uranium.

Due to the surplus of neutrons from Pu-239 fission, the reactor will actually *breed* more Pu-239 than it consumes. The blanket material can then be processed to extract the Pu-239 to replace the losses in the reactor, and the surplus is then mixed with other fuel to produce MOX fuel that can be fed into conventional slow neutron reactors. A single fast reactor can thereby feed several slow ones, greatly increasing the amount of energy extracted from the natural uranium, from less than 1% in a normal once-through cycle, to as much as 60% in the best fast reactor cycles.

Given the limited stores of natural uranium ore, and the rate that nuclear power was expected to take over baseload generation, through the 1960s and 70s fast breeder reactors were seen as the solution to the world's energy needs. Using twice-through processing, a fast breeder economy increases the fuel capacity of known ore deposits by as much as 100 times, meaning that even existing ore sources would last hundreds of years. The disadvantage to this approach is that the breeder reactor has to be fed highly enriched fuel, which is very expensive to produce. Even though it breeds more fuel than it consumes, the resulting MOX is still expensive. It was widely expected that this would still be below the price of enriched uranium as demand increased and known resources dwindled.

Through the 1970s, breeder designs were being widely experimented on, especially in the USA, France and the USSR. However, this coincided with a crash in uranium prices. The expected increased demand led mining companies to build up new supply channels, which came online just as the rate of reactor construction stalled in the mid-1970s. The resulting oversupply caused fuel prices to decline from about US$40 per pound in 1980 to less than $20 by 1984. Breeders produced fuel that was much more expensive, on the order of $100 to $160, and the few units that had reached commercial operation proved to be economically disastrous. Interest in breeder reactors were further muted by Jimmy Carter's April 1977 decision to defer construction of breeders in the US due to proliferation concerns, and the terrible operating record of France's Superphénix reactor.

Advantages

Actinides and fission products by half-life								
Actinides by decay chain				Half-life range (y)	Fission products of ^{235}U by yield			
4*n*	4*n*+1	4*n*+2	4*n*+3			4.5–7%	0.04–1.25%	<0.001%
^{228}Ra$^{№}$				4–6	†		^{155}Eu^{þ}	
^{244}Cm^{f}	^{241}Pu^{f}	^{250}Cf	^{227}Ac$^{№}$	10–29		^{90}Sr	^{85}Kr	^{113m}Cd^{þ}
^{232}U^{f}		^{238}Pu$^{f№}$	^{243}Cm^{f}	29–97		^{137}Cs	^{151}Sm^{þ}	^{121m}Sn
^{248}Bk	^{249}Cf^{f}	^{242m}Am^{f}		141–351	No fission products have a half-life in the range of 100–210 k years ...			
	^{241}Am^{f}		^{251}Cf^{f}	430–900				
		^{226}Ra$^{№}$	^{247}Bk	1.3 k – 1.6 k				
^{240}Pu$^{f№}$	^{229}Th$^{№}$	^{246}Cm^{f}	^{243}Am^{f}	4.7 k – 7.4 k				
	^{245}Cm^{f}	^{250}Cm		8.3 k – 8.5 k				
			^{239}Pu$^{f№}$	24.1 k				
		^{230}Th$^{№}$	^{231}Pa$^{№}$	32 k – 76 k				

$^{236}Np^{f}$	$^{233}U^{f№}$	$^{234}U^{№}$		150 k – 250 k	‡	$^{99}Tc^{¢}$	^{126}Sn	
^{248}Cm		$^{242}Pu^{f}$		327 k – 375 k			$^{79}Se^{¢}$	
				1.53 M		^{93}Zr		
	$^{237}Np^{f№}$			2.1 M – 6.5 M		$^{135}Cs^{¢}$	^{107}Pd	
$^{236}U^{№}$			$^{247}Cm^{f}$	15 M – 24 M			$^{129}I^{¢}$	
$^{244}Pu^{№}$				80 M	... nor beyond 15.7 M years			
$^{232}Th^{№}$		$^{238}U^{№}$	$^{235}U^{f№}$	0.7 G – 14.1 G				

Legend for superscript symbols
¢ has thermal neutron capture cross section in the range of 8–50 barns
f fissile
m metastable isomer
№ naturally occurring radioactive material (NORM)
þ neutron poison (thermal neutron capture cross section greater than 3k barns)
† range 4–97 y: Medium-lived fission product
‡ over 200,000 y: Long-lived fission product

Fast neutron reactors can reduce the total radiotoxicity of nuclear waste, and dramatically reduce the waste's lifetime. They can also use all or almost all of the fuel in the waste. Fast neutrons have an advantage in the transmutation of nuclear waste. With fast neutrons, the ratio between splitting and the capture of neutrons of plutonium or minor actinide is often larger than when the neutrons are slower, at thermal or near-thermal "epithermal" speeds. The transmuted odd-numbered actinides (e.g. from Pu-240 to Pu-241) split more easily. After they split, the actinides become a pair of "fission products." These elements have less total radiotoxicity. Since disposal of the fission products is dominated by the most radiotoxic fission product, cesium-137, which has a half life of 30.1 years, the result is to reduce nuclear waste lifetimes from tens of millennia (from transuranic isotopes) to a few centuries. The processes are not perfect, but the remaining transuranics are reduced from a significant problem to a tiny percentage of the total waste, because most transuranics can be used as fuel.

- Fast reactors technically solve the "fuel shortage" argument against uranium-fueled reactors without assuming unexplored reserves, or extraction from dilute sources such as ordinary granite or the ocean. They permit nuclear fuels to be bred from almost all the actinides, including known, abundant sources of depleted uranium and thorium, and light water reactor wastes. On average, more neutrons per fission are produced from fissions caused by fast neutrons than from those caused by thermal neutrons. This results in a larger surplus of neutrons beyond those required to sustain the chain reaction. These neutrons can be used to produce extra fuel, or to transmute long half-life waste to less troublesome isotopes, such as was done at the Phénix reactor in Marcoule in France, or some can be used for each purpose. Though conventional thermal reactors also produce excess neutrons, fast reactors can produce enough of them to breed more fuel than they consume. Such designs are known as fast breeder reactors.
- The fast reactor doesn't just transmute the inconvenient even-numbered transuranic elements (notably Pu-240 and U-238). It transmutes them, and then fissions them for power, so that these former wastes would actually become valuable.

Disadvantages

- Fast-neutron reactors are costly to build and operate, and are not likely to be cost-competitive with thermal neutron reactors unless the price of uranium increases dramatically.

- Due to the low cross sections of most materials at high neutron energies, critical mass in a fast reactor is much higher than a thermal reactor. In practice, this means significantly higher enrichment: >20% enrichment in a fast reactor compared to <5% enrichment in typical thermal reactors. This raises greater nuclear proliferation and nuclear security issues.
- Sodium is often used as a coolant in fast reactors, because it does not moderate neutron speeds much and has a high heat capacity. However, it burns and foams in air. It has caused difficulties in reactors (e.g. USS Seawolf (SSN-575), Monju), although some sodium-cooled fast reactors have operated safely (notably the Superphénix and EBR-II for 30 years).
- Since liquid metals have low moderating power and ratio and no other moderator is present, the primary interaction of neutrons with liquid metal coolants is the (n,gamma) reaction, which induces radioactivity in the coolant. Boiling in the coolant, e.g. in an accident, would reduce coolant density and thus the absorption rate, such that the reactor has a positive void coefficient, which is dangerous and undesirable from a safety and accident standpoint. This can be avoided with a gas cooled reactor, since voids do not form in such a reactor during an accident; however, activation in the coolant remains a problem. A helium-cooled reactor would avoid this, since the elastic scattering and total cross sections are approximately equal, i.e. there are very few (n,gamma) reactions in the coolant and the low density of helium at typical operating conditions means that the amount neutrons have few interactions with coolant.

Nuclear Reactor Design

Coolant

Water, the most common coolant in thermal reactors, is generally not a feasible coolant for a fast reactor, because it acts as a neutron moderator. However the Generation IV reactor known as the supercritical water reactor with decreased coolant density may reach a hard enough neutron spectrum to be considered a fast reactor. Breeding, which is the primary advantage of fast over thermal reactors, may be accomplished with a thermal, light-water cooled & moderated system using very high enriched (~90%) uranium.

All current fast reactors are liquid metal cooled reactors. The early Clementine reactor used mercury coolant and plutonium metal fuel. Sodium-potassium alloy (NaK) coolant is popular in test reactors due to its low melting point. In addition to its toxicity to humans, mercury has a high cross section for the (n,gamma) reaction, causing activation in the coolant and losing neutrons that could otherwise be absorbed in the fuel, which is why it is no longer used or considered as a coolant in reactors. Molten lead cooling has been used in naval propulsion units as well as some other prototype reactors. All large-scale fast reactors have used molten sodium coolant.

Another proposed fast reactor is a Molten Salt Reactor, one in which the molten salt's moderating properties are insignificant. This is typically achieved by replacing the light metal fluorides (e.g. Lithium fluoride - LiF, Beryllium fluoride - BeF_2) in the salt carrier with heavier metal chlorides (e.g., Potassium chloride - KCI, Rubidium chloride - RbCl, Zirconium chloride - $ZrCl_4$). Moltex Energy based in the UK proposes to build a fast neutron reactor called the Stable Salt Reactor. In this reactor design the nuclear fuel is dissolved in a molten salt. The fuel salt is contained in stainless steel tubes similar to those use in solid fuel reactors. The reactor is cooled using the natural convection of another molten salt coolant. Moltex claims that their design will be less expensive to build than a coal fired power plant and can consume nuclear waste from conventional solid fuel reactors.

Gas-cooled fast reactors have been the subject of research as well, as helium, the most commonly proposed coolant in such a reactor, has small absorption and scattering cross sections, thus preserving the fast neutron spectrum without significant neutron absorption in the coolant.

Nuclear Fuel

In practice, sustaining a fission chain reaction with fast neutrons means using relatively highly enriched uranium or plutonium. The reason for this is that fissile reactions are favored at thermal energies, since the ratio between the Pu239 fission cross section and U238 absorption cross section is ~100 in a thermal spectrum and 8 in a fast spectrum. Fission and absorption cross sections are low for both Pu239 and U238 at high (fast) energies, which means that fast neutrons are likelier to pass through fuel without interacting than thermal neutrons; thus, more fissile material is needed. Therefore it is impossible to build a fast reactor using only natural uranium fuel. However, it is possible to build a fast reactor that will *breed* fuel (from fertile material) by producing more fissile material than it consumes. After the initial fuel charge such a reactor can be refueled by reprocessing. Fission products can be replaced by adding natural or even depleted uranium with no further enrichment required. This is the concept of the fast breeder reactor or FBR.

So far, most fast neutron reactors have used either MOX (mixed oxide) or metal alloy fuel. Soviet fast neutron reactors have been using (high U-235 enriched) uranium fuel. The Indian prototype reactor has been using uranium-carbide fuel.

While criticality at fast energies may be achieved with uranium enriched to 5.5 weight percent Uranium-235, fast reactor designs have often been proposed with enrichments in the range of 20 percent for a variety of reasons, including core lifetime: If a fast reactor were loaded with the minimal critical mass, then the reactor would become subcritical after the first fission had occurred. Rather, an excess of fuel is inserted with reactivity control mechanisms, such that the reactivity control is inserted fully at the beginning of life to bring the reactor from supercritical to critical; as the fuel is depleted, the reactivity control is withdrawn to mitigate the negative reactivity feedback from fuel depletion and fission product poisons. In a fast breeder reactor, the above applies, though the reactivity from fuel depletion is also compensated by the breeding of either Uranium-233 or Plutonium-239 and 241 from Thorium 232 or Uranium 238, respectively.

Control

Like thermal reactors, fast neutron reactors are controlled by keeping the criticality of the reactor reliant on delayed neutrons, with gross control from neutron-absorbing control rods or blades.

They cannot, however, rely on changes to their moderators because there is no moderator. So Doppler broadening in the moderator, which affects thermal neutrons, does not work, nor does a negative void coefficient of the moderator. Both techniques are very common in ordinary light water reactors.

Doppler broadening from the molecular motion of the fuel, from its heat, can provide rapid negative feedback. The molecular movement of the fissionables themselves can tune the fuel's relative speed away from the optimal neutron speed. Thermal expansion of the fuel itself can also provide quick negative feedback. Small reactors such as those used in submarines may use doppler broadening or thermal expansion of neutron reflectors.

Shevchenko BN350 desalination unit. View of the only nuclear-heated desalination unit in the world

History

A 2008 IAEA proposal for a Fast Reactor Knowledge Preservation System notes that:

> during the past 15 years there has been stagnation in the development of fast reactors in the industrialized countries that were involved, earlier, in intensive development of this area. All studies on fast reactors have been stopped in countries such as Germany, Italy, the United Kingdom and the United States of America and the only work being carried out is related to the decommissioning of fast reactors. Many specialists who were involved in the studies and development work in this area in these countries have already retired or are close to retirement. In countries such as France, Japan and the Russian Federation that are still actively pursuing the evolution of fast reactor technology, the situation is aggravated by the lack of young scientists and engineers moving into this branch of nuclear power.

List of Fast Reactors

Fast Reactors of the Past

USA

- CLEMENTINE, the first fast reactor, built in 1946 at Los Alamos National Laboratory. Plutonium metal fuel, mercury coolant, power 25 kW thermal, used for research, especially as a fast neutron source.
- EBR-I at Idaho Falls, which in 1951 became the first reactor to generate significant amounts of electrical power. Decommissioned 1964.
- Fermi 1 near Detroit was a prototype fast breeder reactor that began operating in 1957 and shut down in 1972.

- EBR-II Prototype for the Integral Fast Reactor, 1965–1995.
- SEFOR in Arkansas, a 20 MWt research reactor which operated from 1969 to 1972.
- Fast Flux Test Facility, 400 MWt, Operated flawlessly from 1982 to 1992, at Hanford Washington, now deactivated, liquid sodium is drained with argon backfill under care and maintenance.

Europe

- DFR (Dounreay Fast Reactor, 1959–1977, 14 MWe) and PFR (Prototype Fast Reactor, 1974–1994, 250 MWe), in Caithness, in the Highland area of Scotland.
- Rhapsodie in Cadarache, France, (20 then 40 MW) between 1967 and 1982.
- Superphénix, in France, 1200 MWe, closed in 1997 due to a political decision and very high costs of operation.
- Phénix, 1973, France, 233 MWe, restarted 2003 at 140 MWe for experiments on transmutation of nuclear waste for six years, ceased power generation in March 2009, though it will continue in test operation and to continue research programs by CEA until the end of 2009. Stopped in 2010.
- KNK-II, Germany.

USSR/Russia

- Small lead-cooled fast reactors used for naval propulsion, particularly by the Soviet Navy.
- BR-5 - research fast neutron reactor at the Institute of Physics and Energy in Obninsk. Years of operation 1959-2002.
- BN-350, constructed by the Soviet Union in Shevchenko (today's Aqtau) on the Caspian Sea, 130 MWe plus 80,000 tons of fresh water per day.
- IBR-2 - research fast neutron reactor at the Joint Institute of Nuclear Research in Dubna (near Moscow).
- BN-600 - sodium-cooled fast breeder reactor at the Beloyarsk Nuclear Power Station. Provides 560 MW to the Middle Urals power grid. In operation since 1980.
- BN-800 - sodium-cooled fast breeder reactor at the Beloyarsk Nuclear Power Station. Designed to generate 880 MW of electrical power. Started producing electricity in October, 2014. Achieved full power in August, 2016.

Never operated

- Clinch River Breeder Reactor, USA.
- Integral Fast Reactor, USA. Design emphasized fuel cycle based on on-site electrolytic reprocessing. Cancelled 1994 without construction.

- SNR-300, Germany.
- Monju reactor, 300 MWe, in Japan. was closed in 1995 following a serious sodium leak and fire. It was restarted May 6, 2010 and in August 2010 another accident, involving dropped machinery, shut down the reactor again. As of June 2011, the reactor has only generated electricity for one hour since its first testing two decades prior.

Currently operating

- BN-600, 1981, Russia, 600 MWe, scheduled end of life 2010 but still in operation.
- BN-800, Russia, testing began June 27, 2014, estimated total power 880 MW. Achieved full power in August, 2016.
- BOR-60 - sodium-cooled reactor at the Research Institute of Atomic Reactors in Dmitrovgrad. In operation since 1980.(experimental purposes).
- FBTR, 1985, India, 10.5 MWt (experimental purposes).
- China Experimental Fast Reactor, 65 MWt (experimental purposes), planned 2009, critical 2010.

Under repair

- Jōyō, 1977–1997 and 2004–2007, Japan, 140 MWt. Experimental reactor, operated as an irradiation test facility. After an incident in 2007, the reactor is suspended for repairing, recovery works were planned to be completed in 2014.

Under construction

- PFBR, Kalpakkam, India, 600 MWe.

In design phase

- BN-1200, Russia, build starting after 2014, operation in 2018–2020.
- Toshiba 4S being developed in Japan and was planned to be shipped to Galena, Alaska (USA) but progress is stalled.
- KALIMER, 600 MWe, South Korea, projected 2030. KALIMER is a continuation of the sodium cooled, metallic fueled, fast neutron reactor in a pool represented by the Advanced Burner Reactor (2006), S-PRISM (1998-present), Integral Fast Reactor (1984-1994), and EBR-II (1965-1995).
- Generation IV reactor (Helium·Sodium·Lead cooled) US-proposed international effort, after 2030
- JSFR, Japan, project for a 1500 MWe reactor began in 1998, but without success.
- ASTRID, France, project for a 600 MWe sodium-cooled reactor. Planned experimental operation in 2020.

Planned

- Future FBR, India, 1000 MWe, after 2025.

Chart

Fast reactors				
	U.S.	Russia	Europe	Asia
Past	Clementine, EBR-I/II, SEFOR, FFTF	BN-350	Dounreay, Rhapsodie, Superphé-nix, Phénix (stopped in 2010)	
Cancelled	Clinch River, IFR		SNR-300	
Operating		BOR-60, BN-600, BN-800		FBTR, CEFR
Under repair				Jōyō
Under construc-tion				Monju, PFBR,
Planned	Gen IV (Gas·Sodi-um·Lead)	BN-1200	ASTRID	4S, JSFR, KALIMER

Fast Reactor

Fast reactors generate energy from nuclear fuel through their irradiation with fast neutrons. In a thermal reactor, neutrons produced as a result of neutron absorption in fuel possess high kinetic energy of the order of MeV. These are slowed by elastic collision with moderator resulting in thermal neutrons with energies as low as 0.025 eV. Since the fast reactor utilizes fast neutrons, moderation is not required. To be precise, moderation is undesirable in a fast reactor. Hence fast reactors do not contain moderating materials like water, heavy water and graphite in the core.

The fission cross section of U-235 in fast spectrum is low, compared to that of Pu- 239. Hence Pu-239 is used as the main fissile isotope, though enriched U-235 is used at the start to initiate the chain reaction.

Fast reactors are normally configured for breeding. This requires absorption of neutrons by a blanket of fertile material. Also neutron losses in structural components are to be minimized. Hence layers of blankets containing fertile material are used, to ensure that more fuel is breed than that burnt, qualifying the definition of a breeder.

The most common coolants like water and heavy water cannot be used as coolants in a fast reactor. Non-moderating materials like Helium and liquid metals like sodium, lead, lead-bismuth eutectic qualify to be coolants owing to their non-moderating nature.

Based on the coolant, fast (breeder) reactors are further classified as follows:

(i) sodium cooled fast reactor

(ii) lead cooled fast reactor

(iii) helium cooled fast reactor

Both sodium cooled fast reactor and lead cooled fast reactor are called Liquid Metal Cooled Fast Breeder Reactor (LMFBR).

Due to better transport and neutronic properties, sodium is the most preferred choice for coolant. One of the advantages of using sodium as coolant is the possibility of achieving a high coolant (sodium) outlet temperature, while maintaining a pressure much lower than those maintained for light water and heavy water reactors. This is due to the high boiling point of sodium even at atmospheric pressure. Hence problems associated with high pressures are circumvented to a large extent.

Sodium cooled fast breeder reactors use two cycles of coolant flows. The primary circuit involves the circulation of sodium through the core. Relatively low temperature sodium enters the core at the bottom and leaves at the top at higher temperature. This sodium, called primary sodium is radioactive due to exposure to neutrons while passing through the core.

Another circuit involves heat transfer between the radioactive primary sodium and secondary sodium in separate heat exchangers. The secondary sodium in turn transfers heat to water in steam generator, thus producing steam. The use of secondary coolant between primary coolant and steam is aimed at preventing contact of radioactive sodium with water in case of leakage. While it is to be noted that sodium- water reaction itself is exothermic and needs to be prevented, contact of radioactive sodium with water would also involve concerns with radioactivity. Hence preventing contact between radioactive sodium and water eliminates the radioactivity concerns.

Types of Sodium Cooled Fast Breeder Reactor

There are two types of sodium cooled fast breeder reactor: loop-type reactor, pool- type reactor. This classification is based on the location of (primary) heat exchanger used for transferring heat from primary sodium to secondary sodium.

Loop-type Fast Reactor

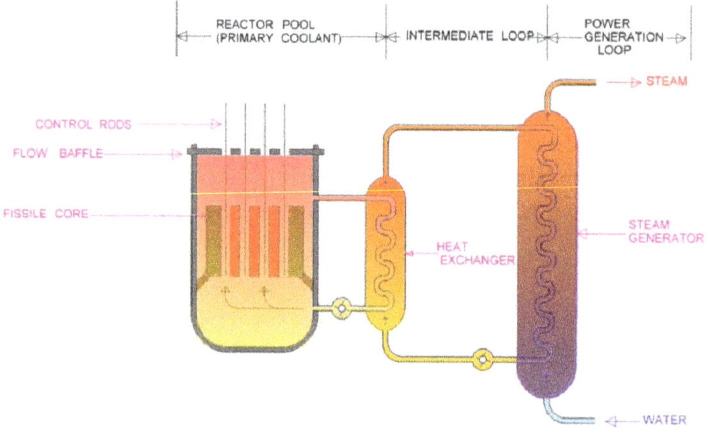

Schematic diagram of loop type fast reactor

The schematic diagram of a loop-type fast reactor is shown in Figure. In loop-type reactor, the heat exchanger used for heat transfer from primary to secondary sodium and the primary sodium pumps is located outside the reactor vessel, but still within the biological shield.

Loop-type Reactor

In pool-type reactor, the primary heat exchanger and pumps for primary sodium are placed inside the reactor tank. As a result, the diameter of reactor tank is higher for the pool design compared to that of loop design.

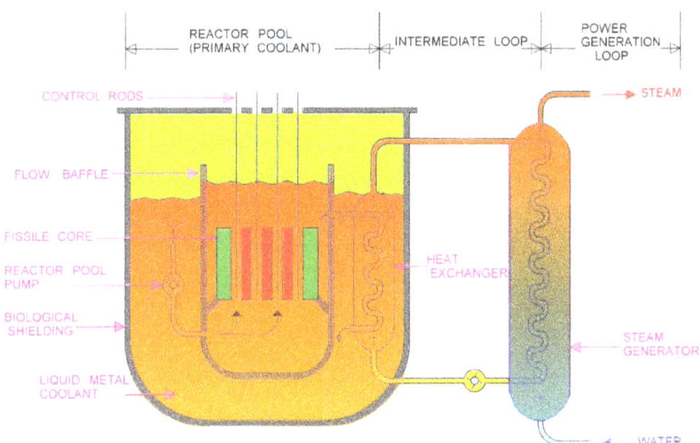

Schematic diagram of pool type fast reactor. Note the presence of heat exchangers inside the large pool of sodium (Redrawn from Ref.)

Also, the thermal inertia of sodium is higher in pool type reactors compared to that in loop-type reactors. For example, when there is a drop in the rate of circulation of primary sodium in the core, the temperature of the coolant due to this transient will slowly increase from the steady-state temperature due to higher heat capacity (product of specific heat and mass). Hence the higher mass of sodium in the pool contributes to the higher thermal inertia.

Other advantages of pool type reactors over loop type reactors are:

(i) Radioactive materials are confined within a single vessel

(ii) Availability of independent and dedicated sodium loop for removing decay heat reliably

(iii) Reduction in the amount of external piping

(iv) No penetrations leading to higher structural integrity of the vessel

(v) Lower neutron dose

(vi) Sodium leakage will not result in Loss of Coolant Accident (LOCA) leading to higher reliability

The core of a fast reactor is very much different from that of a thermal reactor. The key distinction between the core of a fast reactor and that of a thermal reactor is the absence of moderator in fast reactor. It may be recalled that the fast reactor produces energy due to fission caused by fast neutrons (with energy ~ MeV). The presence of a moderator would cause loss of kinetic energy of fast neutrons due to collision with the nuclei of moderator. Hence components made of lighter elements are absent in the fast reactor core.

Another difference between the two cores is their shape. The reactor cores for thermal reactors are rectangular in shape, while a fast (breeder) reactor has a hexagonal core. In thermal reactors, fuel elements are arranged in square lattice. Square lattice provides sufficient space to accommodate water and hence satisfy the required moderator-fuel ratio. With moderator excluded from the core of fast reactors, fuel elements can be arranged closer. With triangular arrangement of fuel elements, higher volume fraction of fuel can be achieved in the core, paving way for reduction in fissile loading. Hence the fast reactor core is hexagonal in shape.

Core Configuration of PFBR

The core configuration is decided taking into account of the breeding requirements and high burnup required for achieving better economics. We shall see the core configuration taking the Prototype Fast Breeder Reactor (PFBR) being built at Kalpakkam as the example. This reactor has been designed with extensive research and operating experience gained by BARC in operating the Fast Breeder Test Reactor at IGCAR, Kalpakkam. It is one among a list of very few fast reactors built across the world.

A schematic diagram of the core of PFBR can be seen in Figure. The core is composed of several subassemblies. The fuel subassembly contains the mixed oxide fuel with axial blanket and shield. Within the fuel subassembly, there are two zones: inner and outer zones. The inner zone (~ 21 % PuO_2) houses 9 control & safety rods and 3 diverse safety rods. The inner zone is surrounded by outer zone (with relatively higher enrichment ~ 28 % PuO_2). The variation in enrichment in the radial direction helps in radial flux flattening.

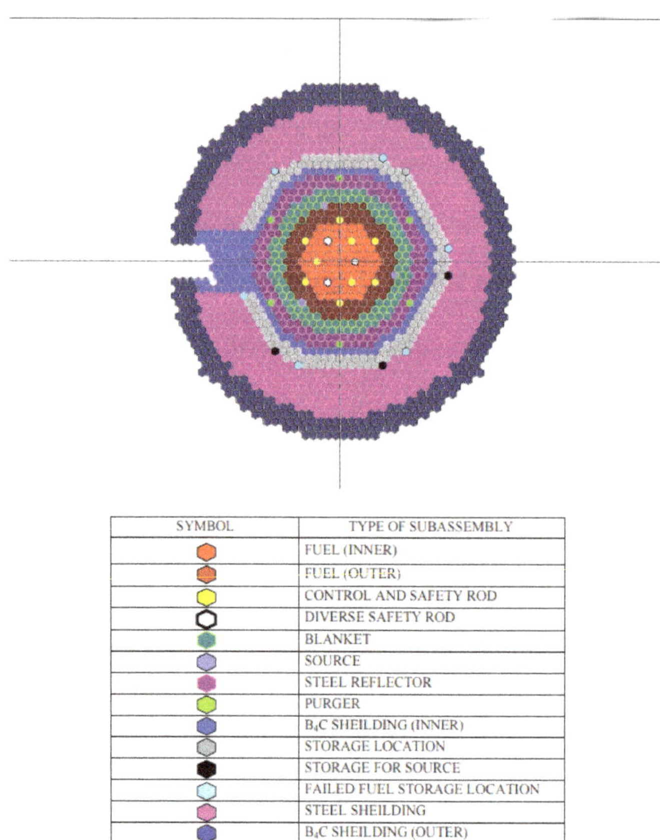

SYMBOL	TYPE OF SUBASSEMBLY
	FUEL (INNER)
	FUEL (OUTER)
	CONTROL AND SAFETY ROD
	DIVERSE SAFETY ROD
	BLANKET
	SOURCE
	STEEL REFLECTOR
	PURGER
	B_4C SHEILDING (INNER)
	STORAGE LOCATION
	STORAGE FOR SOURCE
	FAILED FUEL STORAGE LOCATION
	STEEL SHEILDING
	B_4C SHEILDING (OUTER)

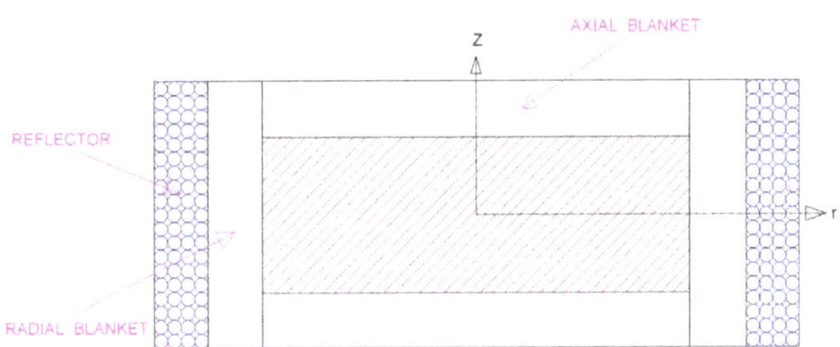

Schematic diagram showing the configuration of PFBR core (Ref.)

In the Prototype Fast Breeder Reactor (PFBR), there are 181 fuel subassemblies with each subassembly containing 217 fuel pins. The pins are of diameter 6 mm. The fuel is mixed oxide (UO_2-PuO_2). The total length of each pin is 1600 mm that comprises 1000 mm of mixed oxide fuel, 300 mm of blanket in the upper and 300 mm of blanket in the lower.

The radial blanket subassembly containing depleted Uranium oxide surrounds the fuel subassembly. Depleted Uranium oxide contains a larger proportion of U-238, a fertile material. Neutrons escaping from the fuel sub assembly encounter depleted Uranium oxide in the radial blanket transmuting them to fissile isotope. This contributes to the generation of fissile material and improves the breeding ratio. This type of fast reactor core is called homogenous core, in which all subassemblies containing the fissile material are located in radial and axial blankets. The fissile and fertile materials are distributed uniformly in the core and hence the term 'homogenous core'.

Stainless steel reflector sub assembly and inner B_4C absorber sub assembly surround the radial blanket subassembly. The spent rods or the fuel rods that have been irradiated for longer duration are moved to the storage location surrounding the inner B_4C sub assembly. Another layer of steel shielding and B_4C shielding (outer) surround the spent rods in storage providing radial shielding.

Reflector is provided to minimize the escape of neutrons out of the reactor core. Also, reflector contributes to flux flattening, which is essential to extract higher power from the fuel without suffering fuel melt down.

Specific Power

It is defined as the reactor power (thermal) per unit mass of fuel present in the reactor. This may be related to the fission cross section (σ_f), number of atoms of fissile isotope per unit mass of the fuel (N_{fm}), average neutron flux (ϕ) as follows:

$$Specific\ power = N_{fm}\phi\sigma_f E_f \qquad (1)$$

From Eq. (1) it is clear that an increase in average neutron flux increases the specific power. The number of fissile isotope per unit mass of the fuel (N_{fm}) depends on the degree of fuel enrichment. Higher levels of enrichment lead to higher specific power.

Power Density

Power density is the amount of thermal energy produced per unit volume of fuel.

The maximum power density of the PFBR is 1763 kW/L while average power density (across the entire core) is 1247 kW/L. The ratio of average flux to maximum flux is 0.71. This ratio is the highest among the prototype fast reactors in the world.

Linear Heat Rating

It is defined as the ratio of thermal power to the product of number of fuel pins and the height (length) of the pin.

The maximum linear rating for PFBR is 450 W/cm with an active core height of 1 m. The linear rating depends on the thermal conductivity of the fuel, surface temperature of the fuel and the centre-line or maximum temperature of the fuel.

Burnup

Burnup is defined as the amount of thermal energy produced (in terms of product of thermal output and number of days) per unit mass of heavy metal. It is the product of specific power in (MW/tonne) and the number of days of reactor operation. Hence, Burnup is expressed in MWd/t. Burnup is an indication of effective utilization of heavy metal with higher burnup indicating extraction of larger quantity of thermal energy from the heavy metal.

The Super-Phenix of France achieved an average burnup of 60,000 MWd/t prior to shutdown. The maximum burnup achieved was 90,000 MWd/t.

The PFBR targets a burnup of 100,000 MWd/t. Higher burnps are preferred due to advantages with respect to the economics of nuclear power. With higher burnups, it is possible to extract more thermal energy and hence electrical energy from the same quantity of fuel. This results in lower expenditure on fuel procurement and fabrication, apart from reduced expenditure in reprocessing. All these bring down the unit cost of electricity generated.

To achieve higher burnups, either higher degree of enrichment or higher neutron flux is required. While the requirement of higher degree of enrichment involves substantial cost, higher neutron flux can be maintained provided appropriate materials are used for various components.

Fluence

Fluence is defined as the integral of neutron flux over the time duration of reactor operation.

$$F = \int_0^t \emptyset dt \qquad (2)$$

This is an indication of neutron dose on reactor materials including structural materials. A fluence of 10^{22} n/cm^2 is expected for FBTR taking the plant life to be 40 years. While choosing the materials for various components of the core and structural components, the net neutron dosage in terms of fluence needs to be considered to evaluate the radiation-induced changes in the material.

Operating Conditions

Let us recall that the Prototype Fast Breeder Reactor has been designed for 1200 MWt and 500 MWe. This gives a plant efficiency of 40 %. The rest constitutes waste heat. The efficiency is dependent on the maximum steam temperature and the minimum temperature of cooling water (coolant) available in condenser. To improve the conversion (thermal-to-electrical) efficiency, steam temperature has to be increased while the coolant temperature has to be maintained at lower values.

Taking into account of material and transient constraints, the steam pressure and temperature have been fixed at 16.7 MPa and 490 °C.

The primary sodium enters at 397 °C and leaves at 547 °C after extracting the fission heat. Secondary sodium enters at 355 °C the reactor and leaves at 525 °C, after extracting the heat from primary sodium. This heat is utilized to generate steam from water that enters at 235 °C.

Choice of Materials

Fuel

The fuel must not undergo phase changes in the operating temperature range of the reactor. As a result, fuel material should possess high melting point and possess high thermal conductivity to facilitate rapid removal of heat. The fuel used must be compatible with coolant and in the case of fast reactors, the fuel must be compatible with liquid sodium in the temperature range of operation. From the economic perspective, easy of fabrication and reprocessing are also important.

Metals, metal oxide and metal carbides can be used as fuel for fast reactors. A comparison of some important characteristics of these fuels is shown in the Table:

Table: Comparison of few important characteristics of different fast reactor fuels

Fuel	Thermal conductivity (W/mK)	Breeding potential
Metal	30-35	1.6
Nitride	12-16	1.3
Carbide	12-16	1.2
Oxide	2-4	1.1

Metallic fuels are bound to swell to a larger extent due to neutron irradiation. This could lead to fuel-clad interaction. At high temperatures, this could cause diffusion of components in the radial direction. An alloy with lower melting point is formed near the clad and affects the structural integrity of clad.

High swelling under neutron irradiation is also the main disadvantage of nitride and carbide fuels.

The thermal conductivity of mixed oxide fuels is lower than that of carbide or metallic counterparts. Also, the thermal conductivity is sensitive to the oxygen to metal atomic ratio. A small deviation of this value from 2 (as per stoichiometry for UO_2 and PuO_2) results in substantial difference in thermal conductivity. And substantial changes in the microstructure occur as a result of irradiation of a fuel. This also results in reduction in the thermal conductivity of the fuel.

The use of oxide fuel requires compromise on breeding ratio and thermal conductivity compared to carbide or metallic fuel. However, oxide fuels are less prone to swelling compared to others and hence to some extent make up for shortcomings in breeding ratio and thermal conductivity. Also the operating experience with mixed oxide fuel is higher compared to that using mixed carbide fuel.

In PFBR, mixed oxide fuel (MOX) is used due to safety of fuel cycle operation and global experience in industrial scale operations, apart from being economically competitive.

The fuel used for fast reactors must have higher degree of enrichment compared to those used in thermal reactors. This is required to compensate for lower fission cross section in fast spectrum compared to that in thermal region. The loss of neutrons from the core for small-sized cores needs to be considered during the decision on degree of enrichment to be utilized.

The enrichment level is calculated as the ratio of Pu-239 to the sum of Pu-239 and U- 235. Different enrichment levels required are obtained by varying the quantity of Pu- 239. Relatively large sized (diameter) reactors like PFBR require an enrichment level of about 0.25 or 25 %, while small reactors require enrichment to the level of 0.7 or 70 %. The degree of enrichment too poses a constraint in the choice of fuel. For example, it is extremely difficult to achieve a degree of enrichment of 0.7 using the metal oxide fuels owing to the requirement of highly enriched uranium. Hence fast reactors with small cores requiring higher degree of enrichment will have to be built with carbide or metallic fuels. This is the case with Indian Fast Breeder Test Reactor (FBTR) operating in Kalpakkam with the mixed carbide (UC-PuC) fuel.

The dimensions of the fuel pins and that of the core are also important. Fast breeder reactors are characterized by relatively lower core height to core diameter ratio. Hence the variation of neutron flux in the axial direction is less pronounced with

ϕ_{max}/ϕ of around 1.2-1.3.

Absorber

Boron carbide is the widely used material for absorber. This is attributed to the high absorption cross section of Boron for neutrons. It is light, brittle and very hard. Boron carbide exists in a wide range of boron composition (B_4C has the lowest boron concentration while $B_{10}C$ has the highest). Another important advantage of Boron carbide is its non-reactivity with sodium, the most preferred coolant for fast reactors. The neutron reaction that occurs during interaction between neutron and boron carbide is

$$^{10}B + {}^{1}n \rightarrow {}^{7}Li + {}^{4}He.$$

Despite the large release of Helium, the crystal structure is not modified.

B_4C must be fabricated as small diameter pellets, with grain size in the sub- micrometer range. Pellets are loaded in pins with vents for release of Helium. The vents facilitate the cooling of pins by sodium.

Coolant

A coolant is a fluid which flows through or around a device to prevent the device from overheating, transferring the heat produced by the device to other devices that either use or dissipate it. An ideal coolant has high thermal capacity, low viscosity, is low-cost, non-toxic, chemically inert, and neither causes nor promotes corrosion of the cooling system. Some applications also require the coolant to be an electrical insulator.

While the term coolant is commonly used in automotive and HVAC applications, in industrial processing heat transfer fluid is one technical term more often used in high temperature as well as low temperature manufacturing applications. The term also covers cutting fluids.

The coolant can either keep its phase and stay liquid or gaseous, or can undergo a phase transition, with the latent heat adding to the cooling efficiency. The latter, when used to achieve below-ambient temperature, is more commonly known as refrigerant.

Gases

Air is a common form of a coolant. Air cooling uses either convective airflow (passive cooling), or a forced circulation using fans.

Hydrogen is used as a high-performance gaseous coolant. Its thermal conductivity is higher than all other gases, it has high specific heat capacity, low density and therefore low viscosity, which is an advantage for rotary machines susceptible to windage losses. Hydrogen-cooled turbogenerators are currently the most common electrical generators in large power plants.

Inert gases are used as coolants in gas-cooled nuclear reactors. Helium has a low tendency to absorb neutrons and become radioactive. Carbon dioxide is used in Magnox and AGR reactors.

Sulfur hexafluoride is used for cooling and insulating of some high-voltage power systems (circuit breakers, switches, some transformers, etc.).

Steam can be used where high specific heat capacity is required in gaseous form and the corrosive properties of hot water are accounted for.

Liquids

The most common coolant is water. Its high heat capacity and low cost makes it a suitable heat-transfer medium. It is usually used with additives, like corrosion inhibitors and antifreeze. Antifreeze, a solution of a suitable organic chemical (most often ethylene glycol, diethylene glycol, or propylene glycol) in water, is used when the water-based coolant has to withstand temperatures below 0 °C, or when its boiling point has to be raised. Betaine is a similar coolant, with the exception that it is made from pure plant juice, and is therefore not toxic or difficult to dispose of ecologically.

Very pure deionized water, due to its relatively low electrical conductivity, is used to cool some electrical equipment, often high-power transmitters and high-power vacuum tubes.

Heavy water is a neutron moderator used in some nuclear reactors; it also has a secondary function as their coolant. Light water reactors, both boiling water and pressurised water reactors the most common type, use ordinary (light) water.

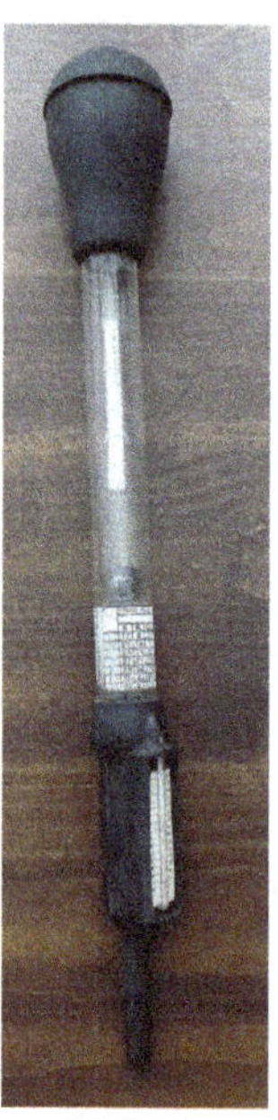

Device to measure the temperature to which the coolant protects the car from freezing.

Polyalkylene glycol (PAG) is used as high temperature, thermally stable heat transfer fluids exhibiting strong resistance to oxidation. Modern PAGs can also be non-toxic and non-hazardous.

Cutting fluid is a coolant that also serves as a lubricant for metal-shaping machine tools.

Oils are used for applications where water is unsuitable. With higher boiling points than water, oils can be raised to considerably higher temperatures (above 100 degrees Celsius) without introducing high pressures within the container or loop system in question.

- Mineral oils serve as both coolants and lubricants in many mechanical gears. Castor oil is also used. Due to their high boiling points, mineral oils are used in portable electric radiator-style space heaters in residential applications, and in closed-loop systems for industrial process heating and cooling. Mineral oil is often used in submerged PC systems as it is non-conductive and therefore won't short circuit or damage any parts.
- Silicone oils and fluorocarbon oils (like fluorinert) are favored for their wide range of operating temperatures. However their high cost limits their applications.
- Transformer oil is used for cooling and additional electric insulation of high-power electric transformers.

Fuels are frequently used as coolants for engines. A cold fuel flows over some parts of the engine, absorbing its waste heat and being preheated before combustion. Kerosene and other jet fuels frequently serve in this role in aviation engines.

Freons were frequently used for immersive cooling of e.g. electronics.

Refrigerants are coolants used for reaching low temperatures by undergoing phase change between liquid and gas. Halomethanes were frequently used, most often R-12 and R-22, but due to environmental concerns are being phased out, often with liquified propane or other haloalkanes like R-134a. Anhydrous ammonia is frequently used in large commercial systems, and sulfur dioxide was used in early mechanical refrigerators. Carbon dioxide (R-744) is used as a working fluid in climate control systems for cars, residential air conditioning, commercial refrigeration, and vending machines.

Heat pipes are a special application of refrigerants.

Molten Metals and Salts

Liquid fusible alloys can be used as coolants in applications where high temperature stability is required, e.g. some fast breeder nuclear reactors. Sodium (in sodium cooled fast reactors) or sodium-potassium alloy NaK are frequently used; in special cases lithium can be employed. Another liquid metal used as a coolant is lead, in e.g. lead cooled fast reactors, or a lead-bismuth alloy. Some early fast neutron reactors used mercury.

For certain applications the stems of automotive poppet valves may be hollow and filled with sodium to improve heat transport and transfer.

For very high temperature applications, e.g. molten salt reactors or very high temperature reactors, molten salts can be used as coolants. One of the possible combinations is the mix of sodium fluoride and sodium tetrafluoroborate ($NaF-NaBF_4$). Other choices are FLiBe and FLiNaK.

Liquid Gases

Liquified gases are used as coolants for cryogenic applications, including cryo-electron microscopy, overclocking of computer processors, applications using superconductors, or extremely sensitive sensors and very low-noise amplifiers.

Carbon Dioxide (chemical formula is CO_2) - is used as a coolant replacement for cutting fluids. CO_2 can provide controlled cooling at the cutting interface such that the cutting tool and the workpiece are held at ambient temperatures. The use of CO_2 greatly extends tool life, and on most materials allows the operation to run faster. This is considered a very environmentally friendly method, especially when compared to the use of petroleum oils as lubricants; parts remain clean and dry which often can eliminate secondary cleaning operations.

Liquid nitrogen, which boils at about -196 °C (77K), is the most common and least expensive coolant in use. Liquid air is used to a lesser extent, due to its liquid oxygen content which makes it prone to cause fire or explosions when in contact with combustible materials.

Lower temperatures can be reached using liquified neon which boils at about -246 °C. The lowest temperatures, used for the most powerful superconducting magnets, are reached using liquid helium.

Liquid hydrogen at -250 to -265 °C can also be used as a coolant. Liquid hydrogen is also used both as a fuel and as a coolant to cool nozzles and combustion chambers of rocket engines.

Nanofluids

A new class of coolants are nanofluids which consist of a carrier liquid, such as water, dispersed with tiny nano-scale particles known as nanoparticles. Purpose-designed nanoparticles of e.g. CuO, alumina, titanium dioxide, carbon nanotubes, silica, or metals (e.g. copper, or silver nanorods) dispersed into the carrier liquid enhance the heat transfer capabilities of the resulting coolant compared to the carrier liquid alone. The enhancement can be theoretically as high as 350%. The experiments however did not prove so high thermal conductivity improvements, but found significant increase of the critical heat flux of the coolants.

Some significant improvements are achievable; e.g. silver nanorods of 55±12 nm diameter and 12.8 µm average length at 0.5 vol.% increased the thermal conductivity of water by 68%, and 0.5 vol.% of silver nanorods increased thermal conductivity of ethylene glycol based coolant by 98%. Alumina nanoparticles at 0.1% can increase the critical heat flux of water by as much as 70%; the particles form rough porous surface on the cooled object, which encourages formation of new bubbles, and their hydrophilic nature then helps pushing them away, hindering the formation of the steam layer. Nanofluid with the concentration more than 5% acts like non-Newtonian fluids.

Solids

In some applications, solid materials are used as coolants. The materials require high energy to vaporize; this energy is then carried away by the vaporized gases. This approach is common in spaceflight, for ablative atmospheric reentry shields and for cooling of rocket engine nozzles. The same approach is also used for fire protection of structures, where ablative coating is applied.

Dry ice and water ice can be also used as coolants, when in direct contact with the structure being cooled.

Sublimation of water ice was used for cooling the space suit for Project Apollo.

Choice of Core Materials

Coolant

The purpose of coolant is to extract the fission heat liberated in the fuel and produce steam for power generation. The coolant ensures that all the materials in the core are below their melting points or below the temperature that causes permanent and serious damage to the functioning of the reactor. Hence coolant plays an important role in maintaining the structural and functional integrity of fuel, cladding and other materials.

The coolant used for fast reactors must be different from the coolant used in thermal reactors. Water and heavy water cannot be used as these are moderating materials that slow down the neutrons. The following criteria are to be borne in mind while choosing an appropriate coolant:

(i) Thermophysical properties

Desirable characteristics: High thermal conductivity, low viscosity, high boiling point, low melting point

(ii) Neutronic properties

Desirable characteristics: Low neutron absorption, low induced radioactivity, negligible moderation

(iii) Material properties

Desirable characteristics: High thermal and radiation stability, compatibility with clad and other structural materials

(iv) Other factors

Desirable characteristics: Affordable cost, non-hazardous

Sodium, lead, lead-bismuth eutectic (LBE) and Helium possess some of the above characteristics. Helium has high specific heat, an advantage for cooling applications. However, its low density requires higher volume, higher pressure and coolant velocity. This compounds already existing problem of lower thermal conductivity of Helium, leading to poor heat transfer rates. Hence, there are no power plants proposed using Helium as the coolant. A detailed discussion on thermophysical properties of various fast reactor coolants will be made in subsequent modules.

For a fixed reactor power, higher coolant velocities and higher neutron fluxes can be used while using sodium as the coolant, when compared to lead and LBE coolants. Also cores of lower diameter with lower enrichment and lower heavy metal inventory can be utilized while operating the reactor with sodium as the coolant. Sodium is compatible with stainless steel also.

In nutshell, compatibility of sodium with stainless steel and its superior thermo- physical and neutronic properties makes sodium as the most preferred choice for fast reactor coolant.

Structural Materials for Core

The important structural materials inside the core are cladding and wrapping materials. These materials experience high neutron dosages, stay in contact with liquid sodium apart from experiencing high temperatures. Hence the choice of material for cladding and wrapping must take into account of irradiation effects, possible corrosion due to contact with sodium and stringent mechanical properties to withstand high temperatures.

Void swelling, irradiation creep and irradiation embrittlement occur due to neutron irradiation. Void swelling refers to increase in the volume of solid material upon prolonged irradiation. In the case of fuel pin, this manifests as increase in volume of pin, though the volume of true solid remains the same. Hence the volume of voids increases, leading to the term "void swelling".

Creep is a material property that represents progressive deformation upon exposure to specific conditions for longer duration of time. Hence, irradiation creep refers to tendency of a material to deform as a result of prolonged irradiation.

Embrittlement refers to loss of a material's ductility. Embrittlement results in brittleness of the material. Such an embrittlement caused by neutron radiation is irradiation embrittlement.

The cladding material is exposed to temperatures in the range of 673-973 K during the course of steady-state operation. Under transient conditions, the temperature may reach 1273 K. Hence the material should possess satisfactory properties in terms of compatibility with fuel, resistance to irradiation, sodium corrosion even at these temperatures.

Taking into account of the above requirements, alloy D9 in 20 % cold worked condition (20 CW D9) is used for cladding. This alloy is superior to 316 SS with respect to void swelling and irradiation creep. For this purpose, the composition of 316 SS is modified by adding Si & Ti in controlled quantities and by increasing the amount of Nickel. Reduction in the concentration of chromium too improves these properties.

20 CW D9 has better mechanical properties and with an inclusion of small corrosion allowance for interaction with sodium, the alloy is satisfactory for cladding and wrapping.

Other Structural Materials

Other structural materials include main vessel, safety vessel, top shields and control rod drive mechanism. Mechanical properties such as tensile strength, creep, creep- fatigue interaction, low cycle and high cycle fatigue must be taken into account.

Low carbon austenitic stainless steel (304 SS & 316 SS) alloyed with nitrogen is used in FBTR. These materials are denoted as 304L(N) and 316L(N). {L stands for low- carbon while N stands for nitrogen addition}

The choice of low carbon content is driven by the fact that sensitization during welding could be avoided with such a material. The reinforcement with 0.06-0.08 wt% nitrogen ensures that the mechanical properties are in par with 304 SS and 316 SS. For components that are likely to experience temperatures above 700 K, 316L(N) is used while for others 304L(N) is used. The technology for production of these materials and fabrication using them are quite mature, justifying their choice.

6

An Integrated Study of Heat Flow Processes

Heat transfer is the interchange of thermal energy between two physical systems. It can be classified into numerous mechanisms like thermal convection, thermal conduction and thermal radiation. Nuclear fission and fusion generate great amounts of heat. The chapter closely examines the key concepts of heat flow process to provide an extensive understanding of the subject.

Heat Generation

The principle modes of heat generation in a nuclear reactor are the reactions (nuclear reactions) that neutrons undergo with nuclei of various materials used in the reactor.

Nuclear fission is exothermic with the energy released due to fission attributed to splitting of heavy nuclei into fragments. The energy released is proportional to the difference in mass of the reactants (neutron & nucleus) and that of products (fission fragments). On an average, 207 MeV (1 MeV = 1.61 x10^{-13} J) of energy is released per fission event in U-235, of which around 200 MeV can be recovered as heat. This energy is sum of the kinetic energies & decay energy of the fission fragments, kinetic energy of new neutrons and the energy of gamma radiation. This energy can be further classified into energy release due to fission and neutron capture.

The recoverable energy released due to fission (200 MeV) can be further classified into instantaneous energy and delayed energy. The components of instantaneous energy release are the kinetic energy of fission fragments (168 MeV), kinetic energy of new neutrons (5 MeV) and gamma radiation (7 MeV). The contributions for the delayed energy come from β-decay (8 MeV) and γ-decay (7 MeV) of fission fragments.

The energy released due to neutron capture is the result of non-fission reactions between excess neutrons (unutilized for chain reaction) and their β-decay and γ-decay (5 MeV).

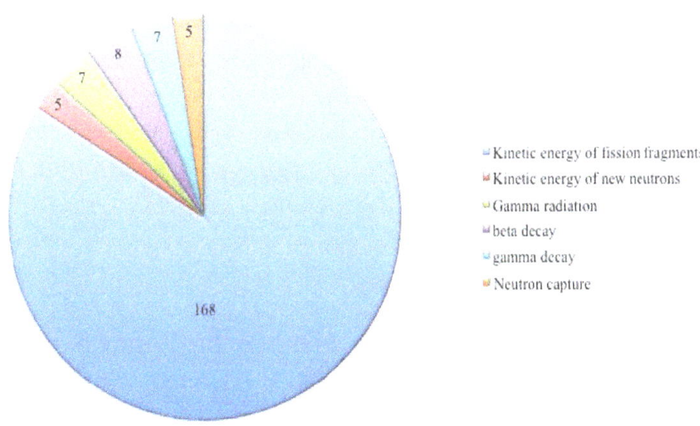

Figure illustrates the energy contribution from a nuclear reactor using U-235 as the fissile isotope. The contributions of various nuclear reactions towards the recoverable energy are also shown.

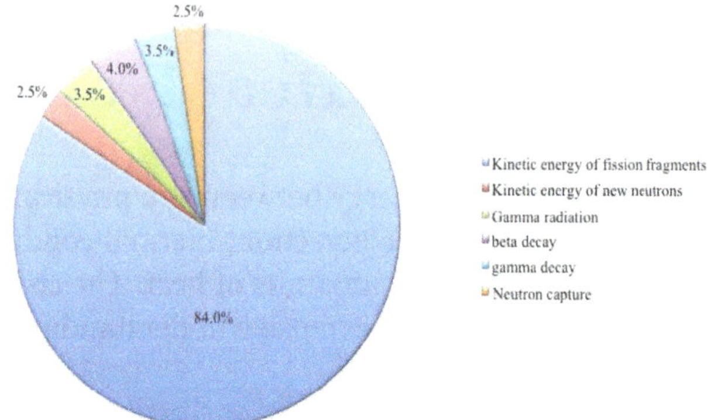

Contribution of different nuclear reactions towards recoverable heat energy: (a) energy in MeV; (b) energy contribution in % of total recoverable energy

Now let's discuss the distribution of this energy among various components of the reactor.

The fission fragments have very short range (< 0.25 mm). The β -radiation too has short range (< 1 mm). Hence their energy is released within the fuel element itself.

The energy released during thermalization of high-energy neutrons rests in the moderator. The energy of the gamma radiation is released in the fuel as well as in structural elements. Table shows the distribution of energy in various components of the reactor.

Tablc: Distribution of fission energy in various components of the thermal reactor

Energy Source	Location of its release
Kinetic energy of fission fragments	Fuel
Kinetic energy of new neutrons	Moderator
γ-radiation (instantaneous)	Fuel and structural components
γ-radiation (delayed)	Fuel and structural components
β-decay of fission products (delayed)	Fuel
Neutron capture (delayed)	Fuel and structural components

For a light-water reactor, 92 % of the total energy released stays in the fuel while about 3 % is released in the moderator. The remaining (5 %) is released in the structural elements.

In a CANDU-type reactor, about 94 % of the total energy is released in fuel, 5% is released in moderator, while the rest is released in pressure tubes, calandria, coolant and shielding.

Reactor Power

Let us look at the factors that influence the power of a reactor. It may be recalled that 200 MeV of (recoverable) energy is released for every fission event (E_f). Hence, the reactor power is the product of energy released per fission (E_f) and the number of fission events.

Energy released (J/s) = Energy released per fission (E_f)* Number of fissions events per second (fission/s)

Neutron flux(ϕ), defined as the number of neutrons per unit fission cross section per unit time is a key parameter in influencing the number of fission events.

Number of fissions events per second (fission/s)= Neutron flux (neutron/cm^{2}s) * Total fission cross section (cm^2).

Note: Please recall that the fission cross section (σ_f) is expressed per nucleus. Hence the total fission cross section is the fission cross section per nucleus multiplied by the number of fuel nuclei.

Number of fissions events per second (fission/s)= Neutron flux (neutron/cm2s)* Fission cross section per nucleus (cm2)*Number of fuel nuclei.

Therefore,

Energy released (J/s) = Energy released per fission (J)* Neutron flux (neutron/cm^{2}s) * Fission cross section per nucleus (cm^2)*Number of fuel nuclei.

Specific power of the fuel, defined as the thermal energy released per unit mass of the fuel, can be related to the energy released as follows:

Specific power of fuel (W/kg) = Energy released (J/s)/Mass of the fuel (kg)

Substituting the above Equations, we get

Specific power of fuel (W/kg) = Energy released per fission (J)* Neutron flux (neutron/cm^{2}s) * Macroscopic fission cross section (cm^2)*Number of fuel nuclei/Mass of the fuel (kg).

Let P' be the specific power of fuel, Ef be the energy released per fission in Joule, N_f be the number of fuel nuclei per unit mass of the fuel then,

$$P'=E_f\phi N_f\sigma_f$$

Equation above is a simplified one assuming neutron flux to be independent of neutron energy and radial position. Taking in to account the influence of radial position and neutron energy on the neutron flux, we have

$$P'=E_fN_f\phi(r,E)\sigma_f$$

Neutron flux corresponding to energy of thermal neutron is the thermal neutron flux, ϕ (r). Hence the specific power of fuel as a function of radial position in the fuel is given by

$$P'=E_fN_f\phi(r)\sigma_f$$

Power density (P") is defined as the thermal energy released per unit volume of the fuel.

$$P''=E_fN_f\phi(r)\sigma_f\rho_f$$

If the average neutron flux is ϕ^-, then the above Eq. can be written as

$$P' = E_f N_f \bar{\phi} \sigma_f$$

$$P'' = E_f N_f \bar{\phi} \sigma_f \rho_f$$

Example – 1: Determine the number of U-235 nuclei in 1 kg of natural UO_2 fuel. The molecular weight of Uranium dioxide is 270.

Solution: From the molecular weight of UO_2 (270) and the atomic weight of Uranium (238), one may calculate the mass of uranium in one kg of UO_2 as follows:

Mass of Uranium in 1 kg of UO_2 = 238/270 = 0.88 kg.

Number of moles of Uranium in 1 kg of UO_2= 0.88/238 = 3.697 gmole.

Recalling the definition of one mole, there are 6.023 x 10^{23} (Avagadro number) atoms or nuclei per mole of a substance.

Therefore, one kg of UO_2 contains 2.227 x 10^{24} atoms.

Please recall that the natural uranium contains only 0.7 % (by mass) of U-235. Hence the number of atoms of U-235 in one kg of UO_2 is the product of atomic fraction of U-235 and 2.227 x 10^{24}.

To convert mass % to atomic fraction, the following method is used.

Atomic % of U-235 = (mass % of U-235/235)/(mass % of U-235/235+ mass % of U- 238/238)

Atomic % of U-235 = (0.7/235)/(0.7/235+99.3/238) = 0.007039.

Note: For natural uranium dioxide fuel, the mass % and atomic % of U-235 are same. However, at higher levels of enrichment the mass % and atomic % may slightly differ.

Therefore, one kg of natural UO_2 contains ~ 2.227 x 10^{24} x 0.007= 1.559 x 10^{22} atoms of U-235.

Example - 2: Determine the specific power and power density of a natural UO_2 fuel in a heavy water reactor. The average neutron flux is 5x10^{13} $cm^{-2}s^{-1}$. The fission cross section is 579 b. The density of UO_2 is 18900 kg/m^3.

Solution:

Writing earlier Eq. again, we have

$$P' = E_f N_f \bar{\phi} \sigma_f$$

E_f= 200 MeV = 3.2 x 10^{-11} J; $\bar{\phi}$=5x10^{13} $cm^{-2}s^{-1}$; σ_f= 579 b = 579*10^{-24} cm^2

As seen in the example -1, the number of fissile nuclei per kg of natural uranium dioxide is 1.559x10^{22}.

Therefore, P' = 14442 W/kg = 14.44 kW/kg

$$P'' = P' \rho_f$$

$$P'' = 273 \text{ MW/m}^3$$

Modes of Heat Transfer

There are three modes by which heat may be transferred from one object to the other:

(i) conduction (ii) convection and (iii) radiation

Conduction

Conduction is a mode of heat transfer that occurs in solids and in stationary fluid. Conduction occurs due to molecular motion in a substance. Molecules in high temperature region possess high thermal energy by virtue of which they move faster. These molecules collide with adjacent molecules causing them to move as well. This causes energy flow towards the region of lower temperature.

Fourier's law of heat conduction can be used to understand the variables influencing heat transfer by conduction. By Fourier's law, rate of heat transfer (Q) per unit area

(A) is proportional to the temperature gradient (dT/dx). The constant of proportionality is thermal conductivity of the material.

$$\frac{Q}{A} = -k\frac{dT}{dx} \qquad (1)$$

'A' refers to the heat transfer area, measured as the area perpendicular to the direction of heat flow. From above Eq, one can understand that to transfer a certain quantity of heat per unit area, a larger temperature gradient is required for a material with lower thermal conductivity. Hence to promote heat transfer by conduction, use of solid materials with higher thermal conductivity is preferred. However, the materials used to minimize heat losses (insulators) must possess lower thermal conductivity.

Example – 1: A plane wall made of a material with thermal conductivity of 1 W/mK is used as a thermal barrier between a hot chamber and atmosphere. The temperature of the wall on the hotter side is 250 °C while that on the colder side is 30 °C. Determine the rate of heat transfer across the wall, if the wall is 1 m x 1 m x 0.1 m.

Solution: The rate of heat transfer can be calculated using Fourier's law of heat conduction as follows:

$$\frac{Q}{A} = -k\frac{dT}{dx}$$

$$Q = -kA\frac{dT}{dx} \qquad (2)$$

k = 1 W/mK; dx = 0.1 m; dT = 220; A = 1 m^2 (Please note that the distance between the hot and cold side is equal to the thickness of the wall)

Substituting the above, in Eq. (2), gives Q = 2200 W

The rate of heat transfer across the wall is 2200 W.

Thermal conductivity is a characteristic property of a material and is also a function of temperature. The thermal conductivity of a material may be related to temperature as:

$$k = k_0(1 + \beta T) \tag{3}$$

For a material with negative value of ' β ' the thermal conductivity of the material

decreases with temperature, while for materials with positive value of ' β ', the thermal conductivity increases with temperature.

When the thermal conductivity of a material is a function of temperature, the rate of heat transfer can be calculated as follows:

$$\frac{Q}{A} = -\frac{1}{L}\int_{T1}^{T2} k(T)dT \qquad ; T_1>T_2 \tag{4}$$

In Eq. (4), 'L' refers to the thickness ($dx = x_2-x_1$)

Convection

Convection is a mode of heat transfer associated with the bulk movement of fluid. When a particle of cold fluid comes into contact with a hot surface, it gains heat from the surface. This fluid particles returns to the bulk and mixes with other particles, during which it transfers heat to the bulk and returns to the bulk temperature. With increased contacts between particles of fluid and the hot surface, more heat is transferred from the hot surface to the fluid. Hence, heat transfer by convection depends on the velocity of fluid and the difference in temperature between the hot surface and the cold fluid. If the movement of fluid is achieved by use of external devices like pumps, stirrers or blowers, it is called forced convection. When fluid motion is set up by density difference existing between the hotter and colder parts of the fluid, it is called natural convection. The rate of heat transfer (Q) per unit heat transfer area (A) is related to the difference in temperature between a surface and liquid as

$$\frac{Q}{A} = h(T_s - T_f)$$

In above Eq, 'h' refers to heat transfer coefficient. Unlike thermal conductivity, heat transfer coefficient is not an inherent property of a material. It is dependent upon the thermo-physical properties of fluid, velocity of the fluid and the geometry of heat transfer system.

Example – 2: A hot cylinder at a temperature of 200 °C is exposed to air at 30°C. If the heat transfer coefficient is 20 W/m²K, determine the rate of heat transfer per unit area.

Solution:

$$\frac{Q}{A} = h(T_s - T_f)$$

h=20 W/m²K; T_s = 200 °C; T_f = 30 °C

Therefore, the rate of heat transfer per unit is 3400 W/m².

Radiation

Radiation is the mode of heat transfer that involves emission and absorption of electromagnetic radiation between two objects placed at different temperatures. This mode of heat transfer does not require a material medium and can occur even in vacuum. This mode of heat transfer is of less importance in nuclear reactors.

Boiling

Boiling is a phase change (liquid to vapour) heat transfer that occurs when a liquid comes into contact with a solid surface or wall at a temperature (T_w) greater than the saturation temperature of the liquid (T_s). Boiling is characterized by high transfer rates achievable due to the high heat transfer coefficient. Hence boiling is used to remove heat from high-heat flux sources, as in nuclear reactors.

Boiling heat transfer may be classified as pool boiling and convective boiling, depending on whether the fluid is at rest or in motion through channels due to pumps. Convective boiling is also called flow boiling.

Let us discuss flow boiling of a liquid flowing inside a vertical heated tube under constant heat flux conditions, as observed for a boiling liquid in the core of a nuclear reactor. As the liquid well below its boiling point enters the heated tube, heat is carried away from the surface by forced convection. The temperature of liquid increases during this phase. This is similar to single phase heat transfer from a heated surface to a flowing liquid. This regime is between A and B in Figure.

When the temperature of liquid bulk is below the saturation temperature, but the surface is at a temperature above the saturation temperature, vapors are produced closer to the surface and are condensed in the liquid bulk. This is called subcooled boiling (boiling when the temperature of liquid bulk is below saturation temperature). This regime is between B and C in Figure. When subcooled boiling takes place in a vertical channel, a two-phase mixture of bubbles and liquid is present in the channel. This is called bubbly flow.

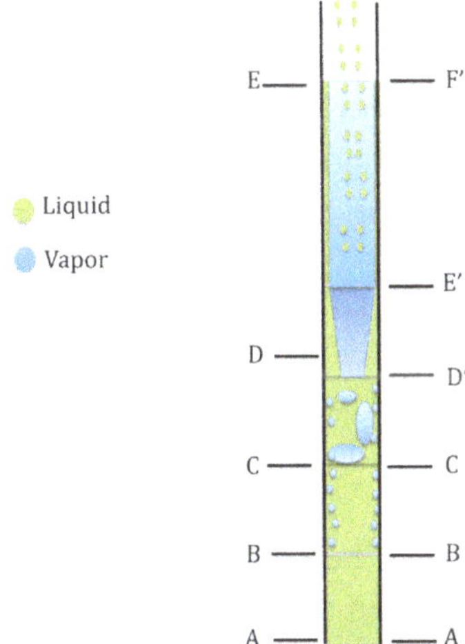

Hydrodynamic and heat transfer regimes for boiling heat transfer

With the fluid moving further upwards in the channel, the temperature of the liquid increases and reaches the saturation temperature. Under these conditions, the vapors formed on the heated surface and that reach the liquid bulk do not condense. Hence the vapors grow as they reach the liquid bulk. The rate of heat removal from the heated surface is very high and hence the temperature of the surface either remains constant or decreases slowly with height. This regime is called saturated nucleate boiling regime (between C and D). The hydrodynamics of this regime is called slug flow (between C and D').

In pressurized water reactors, the vapor fraction is low and hence the surface is cooled by liquid by nucleate boiling. When the vapors begin to blanket the surface, the rate of heat removal is reduced and a condition called "Departure from Nucleate Boiling (DNB)" results. The "Departure from Nucleate Boiling" has to be minimized to ensure that the temperature of clad and fuel do not increase rapidly.

In boiling water reactors, where the fraction of vapors is high, the surface is cooled by the liquid film (similar to that in annular flow). When liquid is completely out of contact with surface, 'Dry out' results. This is also unfavorable as surface temperature increases rapidly.

Neutron Flux

The neutron flux is a quantity used in nuclear reactor physics corresponding to the total length travelled by all neutrons per unit time and volume, or nearly equivalently number of neutrons travelling through a unit area in unit time. The neutron fluence is defined as the neutron flux integrated over a certain time period.

Natural Neutron Flux

Neutron flux in asymptotic giant branch stars and in supernovae is responsible for most of the natural nucleosynthesis producing elements heavier than iron. In stars there is a relatively low neutron flux on the order of 10^5 to 10^{11} neutrons per cm^2 per second, resulting in nucleosynthesis by the s-process (slow-neutron-capture-process). By contrast, after a core-collapse supernova, there is an extremely high neutron flux, on the order of 10^{22} neutrons per cm^2 per second, resulting in nucleosynthesis by the r-process (rapid-neutron-capture-process).

Atmospheric neutron flux, apparently from thunderstorms, can reach levels of $3 \cdot 10^2$ to $5 \cdot 10^2$ neutrons per cm^2 per sec. However, recent results obtained with unshielded scintillation neutron detectors actually show a decrease in the neutron flux during thunderstorms.

Artificial Neutron Flux

Artificial neutron flux refers to neutron flux which is man-made, either as byproducts from weapons or nuclear energy production or for specific application such as from a research reactor or by spallation. A flow of neutrons is often used to initiate the fission of unstable large nuclei. The additional neutron(s) may cause the nucleus to become unstable, causing it to decay (split) to form more stable products. This effect is essential in fission reactors and nuclear weapons.

Within a nuclear fission reactor the neutron flux is primarily the form of measurement used to control the reaction inside. The flux shape is the term applied to the density or relative strength of the flux as it moves around the reactor. Typically the strongest neutron flux occurs in the middle of the reactor core, becoming lower toward the edges. The higher the neutron flux the greater the chance of a nuclear reaction occurring as there are more neutrons going through an area.

A reactor vessel of a typical nuclear power plant (PWR) endures in 40 years (32 full reactor years) of operation approximately 3.5×10^{19} n/cm² (E>1MeV). Neutron flux causes reactor vessels to suffer from embrittlement and the steel gets activated.

Heat Flow Distribution In Plate Fuel Elements

To determine the heat flow distribution, it is essential to study the heat generation in fuel elements as a function of position.

Recalling Eq. (1),

$$\varphi = A\cos\left(\frac{\pi x}{a}\right) = \varphi_{\max}\cos\left(\frac{\pi x}{a}\right) \tag{5}$$

Recalling the relationship between the reactor power per unit volume and neutron flux, we have

$$P'' = E_f N_f \phi(x) \sigma_f \rho_f \tag{6}$$

P" is reactor power per unit volume (W/m³); E_f is the recoverable energy released per fission (J); ϕ (x) is the neutron flux as a function of distance from the centre ($m^{-2}s^{-1}$); N_f is the number of fissile nuclei per unit mass of fuel (kg^{-1}); σ_f is the fission cross section (m²) and ρf is the density of fuel (kgm^{-3})

Substituting Eq. (5) in Eq. (6) we get,

$$P'' = E_f N_f \varphi_{\max}\cos\left(\frac{\pi x}{a}\right)\sigma_f \rho_f \tag{7}$$

Equation (7) may be simplified as:

$$P'' = K'\cos\left(\frac{\pi x}{a}\right) \tag{8}$$

where K' = $\phi_{max} E_f N_f \sigma_f \rho_f$

Equation (8) relates the heat generated as a function of distance from the centre of the slab. Equation (1) is similar to Eq. (8) with K' of Eq. (8) replaced by ϕ_{max} of Eq. (5). Hence profile of reactor power as a function of distance from the centre will be qualitatively similar to neutron flux profiles as shown in Figure.

Temperature Distribution in Plate Fuel Elements

The temperature profile in the slab reactor can be obtained by the solution of steady- state energy balance equation with heat generation.

The state-state energy balance equation in Cartesian coordinates with P" as the heat generation rate per unit volume is given by

$$\frac{\partial^2 T}{\partial x^2}+\frac{\partial^2 T}{\partial y^2}+\frac{\partial^2 T}{\partial z^2}+\frac{P''}{k}=0 \tag{9}$$

Applying Eq. (9) for an infinite slab reactor as one-dimensional steady-state heat conduction, we have

$$\frac{\partial^2 T}{\partial x^2}+\frac{P''}{k}=0 \tag{10}$$

Substituting Eq. (8) in Eq. (10),

$$\frac{d^2 T}{dx^2}+\frac{K'\cos\left(\frac{\pi x}{a}\right)}{k}=0 \tag{11}$$

Two boundary conditions are required for the solution of Eq. (11).

At the centre of the slab, neutron flux and hence heat generation are maximum. Hence, one would expect the temperature at the centre of the reactor to be the maximum.

Therefore,

$$x=0; \quad \frac{dT}{dx}=0 \tag{12}$$

The other boundary condition is the temperature at the outer surface of the reactor. Let it be T_S.

Therefore,

$$x=\frac{a}{2}; \quad T=T_S \tag{13}$$

Solution of Eq. (11) subjected to the boundary conditions (Eq. (12) & Eq. (13)) leads to

$$T=T_S+\frac{K'a^2}{k\pi^2}\left(\cos\frac{\pi x}{a}\right) \tag{14}$$

If the heat generated is assumed to be constant with position, or if the average heat generated is taken into account, then Eq. (10) becomes

$$\frac{d^2 T}{dx^2}+\frac{P''_{avg}}{k}=0 \tag{15}$$

The solution of Eq. (15) subjected to the boundary conditions (Eq. (12) & Eq. (13)) is

$$T = T_S + \frac{P''_{avg}}{8k}\left(a^2 - x^2\right) \quad (16)$$

Example 1: A slab of 100 mm thick made up of a material of thermal conductivity 120 W/mK generates heat at a rate of 1.5 MW/m³. Determine the temperature at the centre of the slab, if the outer surface of the slab is maintained at 180 °C.

Data: k = 120 W/mK; T_S = 180 °C; P_{avg}"=1.5x 10^6 W/m³; a = 0.10 m

Substituting above and x= 0 in Eq. (16), we get T = 196 °C

Example 2: A slab of 100 mm thick made up of a material of thermal conductivity 120 W/mK generates heat at a rate of 1.5cos(πx/a) MW/m³, where 'x' is the distance from the centre of the slab and 'a' is thickness of the slab. Determine the temperature at the centre of the slab, if the outer surface of the slab is maintained at 180 °C.

Data: k = 120 W/mK; T_S = 180 °C; P"=1.5x 106cos(π x/a); K'= 1.5 x 10^6 W/m³; a = 0.10 m

Equation (14) is to be used for this purpose:

$$T = T_S + \frac{K'a^2}{k\pi^2}\left(\cos\frac{\pi x}{a}\right)$$

Solving the above equation for x = 0 gives T= 192.7°C

Equations (14) and (16) can be used to predict the temperature profiles in a slab reactor only if T_S is known. In many instances T_S is not known.

Let us consider the infinite slab to be cooled by water as in the case of a water-cooled reactor. Under these circumstances, the temperature profile in the slab reactor along with its surface temperature can be predicted by equating the heat generated in the slab to the heat removed by coolant from the slab.

Heat generated in the slab per unit volume = P_{avg}"

Total heat generated in the slab = P_{avg}"a LB (17)

In Eq. (17), L and B are the length and breadth of the slab, such that their product is the area perpendicular to the direction of heat flow.

One may recollect the relationship between heat gained by the coolant, heat transfer coefficient and the driving force as

Heat gained by the coolant = hAΔT (18)

In Eq. (18), 'h' is the heat transfer coefficient, 'A' the heat transfer area (LB) and ΔT, the difference in temperature between hot surface (T_S) and the coolant (T_c).

Therefore,

$P_{avg}''a\ LB = h\ A\ (T_S - T_c)$ (19)

Since 'A=LB', we have

$P_{avg}''a = h\ (T_S - T_c)$ (20)

Therefore,

$$T_S = T_c + \frac{P''_{avg}a}{h} \quad (21)$$

Example - 3: A slab of thickness 200 mm generates heat at the rate of 200 kW/m³. If the slab is cooled by water at a temperature of 25 °C, determine the temperature on the outer surface of the slab. The heat transfer coefficient may be taken as 400 W/m²K.

Data: T_c = 25 °C, h = 400 W/m²K; $P_{avg}''=2x10^5$ W/m³; a = 0.2 m

Substituting the above in Eq. (21), we get T_S = 125 °C.

Power Profile in Finite Cylindrical Reactor

Please recall the neutron flux profile for a cylindrical reactor

$$\phi = \phi_0 J_0\left(\frac{2.405r}{R'}\right)\cos\left(\frac{\pi z}{H'}\right) \quad (22)$$

Recalling the relationship between reactor power per unit volume (P") and neutron flux (ϕ), we have

$$P'' = E_f N_f \phi(z,r) \sigma_f \rho_f \quad (23)$$

Substituting Eq. (22) in Eq. (23) we get,

$$P'' = E_f N_f \sigma_f \rho_f \phi_0 J_0(2.405\frac{r}{R'})\cos(\pi\frac{z}{H'}) \quad (24)$$

Equation (24) may be simplified as:

$$P'' = K'' J_0(2.405\frac{r}{R'})\cos(\pi\frac{z}{H'}) \quad (25)$$

where K" = $E_f N_f \sigma_f \rho_f$

Equation (25) relates the heat generated as a function of distance from the centre of the cylinder and as function of distance from the bottom. Equation (22) is similar to Eq. (25) with K" of Eq. (25) replaced by ϕ_0 in Eq. (22). Hence profile of reactor power as a function of distance from the centre will be qualitatively similar to neutron flux profiles as shown in Figure.

Temperature Distribution in Cylindrical Fuel Elements

Analogous to the prediction of temperature profile for an infinite slab reactor, the temperature profile in cylindrical fuel elements can also be obtained by the solution of one-dimensional, steady-state energy balance equation with heat generation in cylindrical coordinates. This is shown as Equation (26)

$$\frac{1}{r}\frac{d}{dr}\left(r\frac{dT}{dr}\right)+\frac{P''}{k}=0 \tag{26}$$

To begin with, let us solve Eq. (26) for the case of uniform volumetric rate of heat generation, P''_{avg}

Equation (26) becomes

$$\frac{1}{r}\frac{d}{dr}\left(r\frac{dT}{dr}\right)+\frac{P''_{avg}}{k}=0 \tag{27}$$

The temperature will be maximum at the centre of the fuel element. Hence one of the boundary conditions is

At r = 0; dT/dr=0 (28)

Let the temperature on the outer surface of the fuel be T_S. Accordingly, the second boundary condition is

At r = R; T = T_S (29)

To solve Eq. (27) using the boundary conditions given in Eq. (28) and Eq. (29), we may rearrange Eq. (27) as follows:

$$\frac{d}{dr}\left(r\frac{dT}{dr}\right)+\frac{rP''_{avg}}{k}=0 \tag{30}$$

Integrating Eq. (30),

$$r\left(\frac{dT}{dr}\right)=\frac{-r^2P''_{avg}}{2k}+C_1 \tag{31}$$

Applying the first boundary condition (r =0; dT/dr = 0) gives C_1 = 0 and Eq. (31) becomes

$$r\left(\frac{dT}{dr}\right)=\frac{-r^2P''_{avg}}{2k} \tag{32}$$

Integrating Eq. (32), we get

$$T=\frac{-r^2P''_{avg}}{4k}+C_2 \tag{33}$$

Applying the second boundary condition (r = R; T = T_S), we get

$$C_2 = T_S + \frac{R^2 P''_{avg}}{4k} \quad (34)$$

Substitution of Eq. (34) in Eq. (27) gives

$$T = T_S + \frac{R^2 P''_{avg}}{4k} - \frac{r^2 P''_{avg}}{4k} \quad (35)$$

Re-arranging Eq. (35),

$$T = T_S + \frac{R^2 P''_{avg}}{4k}\left(1 - \frac{r^2}{R^2}\right) \quad (36)$$

The surface temperature (at the outer surface of the cylinder) can be obtained by equating the heat generated in the cylinder with the heat removed by the coolant as follows (similar to how it was done for slab geometry)

$$P''_{avg} H \frac{\pi D^2}{4} = h\pi DH(T_S - T_c) \quad (37)$$

Note: π DH is the heat transfer area or the area perpendicular to the direction of heat transfer. $\pi D^2/4$ is the cross sectional area which when multiplied by the height of the cylinder (H) gives its volume. T_c is the temperature of the coolant.

Rearrangement of Eq. (37) gives

$$T_S = T_c + P''_{avg} \frac{D}{4h} \quad (38)$$

The maximum temperature (centre-line temperature) is obtained by substituting r=0 and Eq. (38) in Eq. (36) as follows:

$$T = T_c + P''_{avg} \frac{D}{4h} + \frac{R^2 P''_{avg}}{4k} \quad (39)$$

Example – 1: A cylindrical fuel rod of diameter 1.1 cm generates heat at the rate of 10 MW/m^3. If water at a temperature of 220 °C extracts the heat from the fuel rod, determine the maximum temperature of the fuel rod and the surface temperature of the fuel rod. The heat transfer coefficient may be taken as 1000 W/m^{2}K. The thermal conductivity of fuel is 30 W/mK.

Data: T_c = 220°C; $P_{avg}''=10^7$ W/m^3; D = 0.011 m; h = 1000 W/m^{2}K; k = 30 W/mK

Before estimating the maximum temperature, the temperature on the outer surface of the rod (T_S) can be calculated using Eq. (38)

T_S = 247.5 °C

Now, Eq. (15) can be used with r = 0 and T_S=247.5 to determine the maximum temperature as

T_{max} =250 °C.

(Note: The lower temperature difference between the temperature at the centre of the rod and that at the outer surface is low due to smaller diameter of the rod).

Temperature Distribution in Cylindrical Fuel Elements with Position Dependent flux

Recall Eq. (24)

$$P'' = E_f N_f \sigma_f \rho_f \phi_0 J_0(2.405\frac{r}{R'})\cos(\pi\frac{z}{H'})$$

At a fixed axial position, P" is dependent on 'r' alone may be expressed as

$$P'' = K'''r/R = P_{max}''r/R \tag{40}$$

Substituting Eq. (40) in Eq. (27), we have

$$\frac{1}{r}\frac{d}{dr}\left(r\frac{dT}{dr}\right)+\frac{P''_{max}r}{kR}=0 \tag{41}$$

Eq. (41) can be solved as follows:

$$\frac{d}{dr}\left(r\frac{dT}{dr}\right)+\frac{P''_{max}r^2}{kR}=0 \tag{42}$$

Integrating Eq. (42), we get

$$\left(r\frac{dT}{dr}\right)=-\frac{P''_{max}r^3}{3kR}+C \tag{43}$$

Applying the first boundary condition (r =0; dT/dr=0),

$$C=0 \tag{44}$$

Therefore, Eq. (43) becomes

$$\left(\frac{dT}{dr}\right)=-\frac{P''_{max}r^2}{3kR} \tag{45}$$

Integrating Eq. (45),

$$T = -\frac{P''_{max} r^3}{9kR} + C' \tag{46}$$

Applying the second boundary condition (r= R; T = T_s)

$$C' = T_s + \frac{P''_{max} R^3}{9kR} \tag{47}$$

Substituting Eq. (47) in Eq. (46),

$$T = T_s + \frac{P''_{max} R^3}{9kR} - \frac{P''_{max} r^3}{9kR} \tag{48}$$

$$T = T_s + \frac{P''_{max} R^3}{9kR}\left(1 - \frac{r^3}{R^3}\right) \tag{49}$$

Equation (38) holds good for the determination of 'T_s'

Temperature Distribution in Cylindrical Fuel Elements with Temperature Dependent Thermal Conductivity and Uniform Heat Generation

So far, we have assumed thermal conductivity to be independent of temperature. However, thermal conductivity of solids is a function of temperature.

Recalling Eq. (24),

$$k = k_0(1 + \beta T) \tag{50}$$

Recalling Eq. (26),

$$\frac{1}{r}\frac{d}{dr}\left(r\frac{dT}{dr}\right) + \frac{P''}{k} = 0 \tag{51}$$

Substituting Eq. (50) in Eq. (51) results in

$$\frac{1}{r}\frac{d}{dr}\left(r\frac{dT}{dr}\right) + \frac{P''}{k_0(1+\beta T)} = 0 \tag{52}$$

Rearranging Eq. (52),

$$\frac{d}{dr}\left(r\frac{dT}{dr}\right) + \frac{rP''}{k_0(1+\beta T)} = 0 \tag{53}$$

Integrating Eq. (53),

$$\left(r\frac{dT}{dr}\right) = -\frac{r^2 P''}{2k_0(1+\beta T)} + C \qquad (54)$$

At r = 0; dT/dr = 0

Therefore, C = 0 and Eq. (54) becomes

$$\left(\frac{dT}{dr}\right) = -\frac{rP''}{2k_0(1+\beta T)} \qquad (55)$$

Integrating Eq. (55),

$$\int (1+\beta T)dT = -\int \frac{rP''}{2k_0} dr \qquad (56)$$

$$T + \frac{\beta T^2}{2} = -\frac{r^2 P''}{4k_0} + C' \qquad (57)$$

Substituting the second boundary condition, (r = R; T = T_s)\

$$C' = T_s + \frac{\beta T_s^2}{2} + \frac{R^2 P''}{4k_0} \qquad (58)$$

Substitution of Eq. (58) in Eq. (57) gives,

$$T + \frac{\beta T^2}{2} = T_s + \frac{\beta T_s^2}{2} + \frac{R^2 P''}{4k_0}\left(1 - \frac{r^2}{R^2}\right) \qquad (59)$$

Equation (59) is the expression for temperature as a function of radial position. If the thermal conductivity is taken to be independent of temperature, β =0. Substituting β =0 in Eq. (59) gives,

$$T = T_s + \frac{R^2 P''}{4k_0}\left(1 - \frac{r^2}{R^2}\right) \qquad (60)$$

Please note that Eq. (60) is same as that of Eq. (57)

Equation (59) is a quadratic equation, solution of which is given below:

$$T = \frac{-1 + \sqrt{1 + 2\beta\left\{T_S + \beta\frac{T_S^2}{2} + \frac{P''R^2}{4k_0}\left(1 - \frac{r^2}{R^2}\right)\right\}}}{\beta} \qquad (61)$$

In which temperature distribution within a cylindrical fuel element was studied. The analysis of temperature distribution will include cladding also.

The learners will be able to estimate the temperature distribution in cylindrical fuel element and cladding through steady-state energy balance.

It may be recalled that the cladding is used to prevent direct contact between the coolant and fuel. Hence cladding presents an additional resistance for heat transfer between coolant and fuel. Also a small gap exists between fuel and cladding called fuel-clad gap for accommodation of fission gases.

Let us analyze the temperature distribution in cylindrical fuel and the cladding using steady state energy balance in cylindrical coordinates. Figure shows the geometry considered for the analysis.

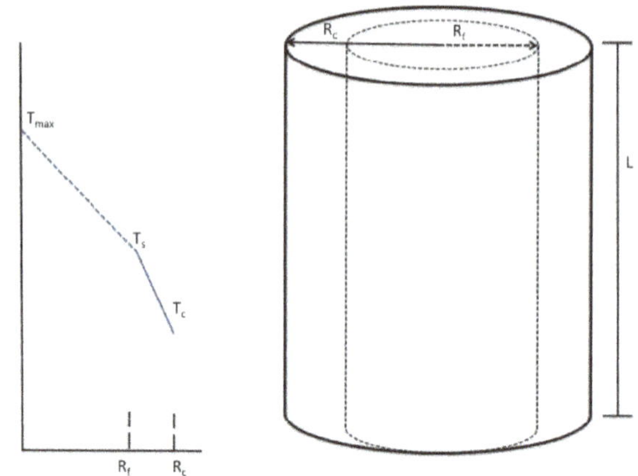

Geometry of cylindrical fuel element with cladding for analysis

Nomenclature

R_f, R_c – Radii of fuel and cladding respectively

T_f, T_c – Temperature on the outer surface of fuel and cladding respectively

T_{max} – Maximum temperature (at the centre of the fuel)

T_∞ - Temperature of the coolant

k_f, k_c – Thermal conductivities of fuel and cladding material respectively

Determination of T_c

In Figure, there are three unknown temperatures: T_{max}, T_f and T_c. It is easier to calculate the T_c, the temperature at the outer surface of the cladding by making a simple energy balance.

Under steady-state conditions, the rate of heat transfer is same across the fuel layer and cladding layer. This is also same as the rate of heat transfer from cladding to the coolant.

Heat transfer rate from cladding to the coolant = $h(T_c - T_\infty)\, \pi\, R_c L$ (62)

If P" is the volumetric rate of heat generation in the fuel, then the rate of heat generation for the entire fuel volume is 'Q'

$$Q = P''\pi R_f^2 L \quad (63)$$

This rate of heat generation in the fuel has to be ultimately removed by the coolant. Hence Equating (62) & (63),

$$Q = P''\pi R_f^2 L = h(T_c - T_\infty)\pi R_c L \tag{64}$$

Therefore,

$$T_c = T_\infty + \frac{P''R_f^2}{hR_c} \tag{65}$$

Equation (65) can be used to determine the temperature on the outer surface of the cladding.

Determination of T_s

The rate of heat transfer across any surface is the ratio of driving force to heat transfer resistance (R_T).

Accordingly, heat transfer across cladding layer is equal to ratio of temperature difference across the cladding layer to the heat transfer resistance in cladding layer.

Recalling Fourier's law of heat conduction,

$$\frac{Q}{A} = -k\frac{dT}{dx} \tag{66}$$

Rewriting Eq. (66) for an elemental cylindrical volume of Length (L) and radius (r), we have

$$Q = -k(2\pi rL)\frac{dT}{dr} \tag{67}$$

Eq. (67) can be integrated with the following limits:

At r = R_f, T = T_f;

&

At r = R_c, T= T_c;

The solution is shown below:

$$Q = \frac{2\pi Lk_c(T_f - T_c)}{ln\left(\frac{R_c}{R_f}\right)} \tag{68}$$

Rearranging Eq. (68), we have

$$T_f = T_c + \frac{Qln\left(\frac{R_c}{R_f}\right)}{2\pi Lk_c} \tag{69}$$

Temperature (T) at any location (r) in the fuel can be determined as follows, from the knowledge of temperature of the fuel (T_f) on its outer surface (R_f):

$$T = T_f + \frac{R_f^2 P''}{4k_f}\left(1 - \frac{r^2}{R_f^2}\right) \tag{70}$$

Substituting Eq. (69) in Eq. (70),

$$T = T_c + \frac{Q\ln(R_c/R_f)}{2\pi L k_c} + \frac{R_f^2 P''}{4k_f}\left(1 - \frac{r^2}{R_f^2}\right) \tag{71}$$

Equation (71) can be used to predict temperature profile in the cylindrical fuel element.

Substituting r=0 in Eq. (71), one can determine the maximum fuel temperature as given below:

$$T_{max} = T_c + \frac{Q\ln(R_c/R_f)}{2\pi L k_c} + \frac{R_f^2 P''}{4k_f} \tag{72}$$

Example – 1: A cylindrical fuel rod of 8 mm radius made of a material with thermal conductivity 0.7 W/mK generates heat at the rate of 1000 kW/m³. The fuel rod is surrounded by a cladding layer of thickness 4 mm. The thermal conductivity of cladding layer is 1.5 W/mK. If the cladding is cooled by coolant at 30 °C, determine the maximum temperature in the fuel, temperature at the outer surface of the fuel and temperature at the outer surface of the cladding. Heat transfer coefficient may be taken as 900 W/m²K. The length of fuel rod is 1 m.

Data:

$R_f = 8\text{x}10^{-3}$ m; $R_c = R_f$+clad thickness = 12×10^{-3} m

$k_f = 0.7$ W/mK; $k_c = 1.5$ W/mK

h = 900 W/m²K

$T_\infty = 30$ °C

P" = 1×10^6 W/m³

Step – I: Determine T_c using Eq. (4)

$$T_c = T_\infty + \frac{P'' R_f^2}{h R_c}$$

$T_c = 30 + 1e6*(8\times10^{-3})^2/(900*12\times10^{-3}) = 35.93$ °C

Step – II: Determine Q

$Q = P''\pi R_f^2 L = 1\times10^6 * 3.14*(8\times10^{-3})^2*1 = 200.96$ W

Step – III: Determine T_f using Eq. (69)

$$T_f = T_c + \frac{Q\ln\left(\frac{R_c}{R_f}\right)}{2\pi L k_c}$$

$$T_f = 35.93 + \frac{200.96\ln(12/8)}{2\pi*1.5} = 44.58$$

Step – IV: Determine maximum fuel temperature (T_{max}) from Eq. (72)

$$T_{max} = T_c + \frac{Q\ln(R_c/R_f)}{2\pi L k_c} + \frac{R_f^2 P''}{4k_f}$$

T_{max} = 67.44 °C

Temperature Distribution in Cylindrical Fuel Elements with Position Dependent flux

The volumetric rate of heat generation per unit volume of fuel is related to radial position in the fuel as follows:

$P'' = K'''r/R_f = P_{max}''r/R_f$ (73)

T_c has to be determined to begin with. However, calculation of T_c requires heat transfer rate, which in turn requires average volumetric rate of heat generation.

The average volumetric rate of heat generation P_{avg}'' can be determined as follows:

$$P''_{avg} = \frac{1}{\pi R_f^2}\int_0^{R_f} P''\, 2\pi r dr = \frac{1}{\pi R_f^2}\int_0^{R_f} P''_{max}\, \frac{r}{R_f} 2\pi r dr \quad (74)$$

$$P''_{avg} = \frac{2P''_{max}}{R_f^3}\int_0^{R_f} r^2 dr = \frac{2P''_{max}}{R_f^3}\left(\frac{R_f^3}{3}\right) \quad (75)$$

Therefore,

$P_{avg}'' = 2P_{max}''/3$ (76)

Replacing P" in Eq. (65) with P_{avg}'', we get

$$T_c = T_\infty + \frac{P''_{avg} R_f^2}{hR_c} \quad (77)$$

Rate of heat transfer becomes

$$Q = P_{avg}''\, \pi\, R_f^2 L \quad (78)$$

The temperature on the surface of the fuel rod can be determined using Eq. (69). The maximum temperature at the centre of the fuel can be determined using Eq. (72) by replacing (P") of Eq. (72) and P_{avg}'' as follows:

$$T_{max} = T_c + \frac{Q\ln(R_c/R_f)}{2\pi L k_c} + \frac{R_f^2 P''_{avg}}{4k_f} \quad (79)$$

Example – 2: A cylindrical fuel rod of 20 mm radius made of a material with thermal conductivity 1 W/mK generates heat at the rate of 3000(r/R_f) kW/m³. The fuel rod is surrounded by a cladding layer of thickness 5 mm. The thermal conductivity of cladding layer is 3 W/mK. If the cladding is cooled by coolant at 25 °C, determine the maximum temperature in the fuel, temperature at the outer surface of the fuel and temperature at the outer surface of the cladding. Heat transfer coefficient may be taken as 1000 W/m²K. The length of fuel rod is 1 m.

Step – I: Determine average volumetric rate of heat generation using Eq. (76)

$P_{avg}" = 2P_{max}"/3 = 2000$ kW/m³

Using Eq. (16), T_c can be calculated as follows:

$$T_c = T_\infty + \frac{P''_{avg} R_f^2}{h R_c}$$

$T_c = 57$ °C

Rate of heat transfer (Q) can be found out using Eq. (17)

$Q = Pavg" \pi R_f^2 L$ = 2000e3*3.14*(20e-3)^2*1 = 2512 W

Using Eq. (8), T_f can be calculated as shown below:

$$T_f = T_c + \frac{Q \ln\left(\frac{R_c}{R_f}\right)}{2\pi L k_c}$$

$T_f = 86.75$ °C

Using Eq. (79), the maximum temperature at the centre of the fuel rod can be calculated as:

$$T_{max} = T_c + \frac{Q\ln(R_c/R_f)}{2\pi L k_c} + \frac{R_f^2 P''_{avg}}{4k_f};$$

$T_{max} = 286.75$ °C

Steam Generator (Nuclear Power)

The inverted U-tube bundle of a Combustion Engineering steam generator.

Steam generators are heat exchangers used to convert water into steam from heat produced in a nuclear reactor core. They are used in pressurized water reactors (PWR) between the primary and secondary coolant loops.

In other types of reactors, such as the pressurised heavy water reactors of the CANDU design, the primary fluid is heavy water. Liquid metal cooled reactors such as the Russian BN-600 reactor also use heat exchangers between primary metal coolant and at the secondary water coolant.

Boiling water reactors (BWR) do not use steam generators, as turbine steam is produced directly in the reactor core. Activation of oxygen and dissolved nitrogen in the water means that the turbine hall is inaccessible during reactor operation and for some time afterwards.

Description

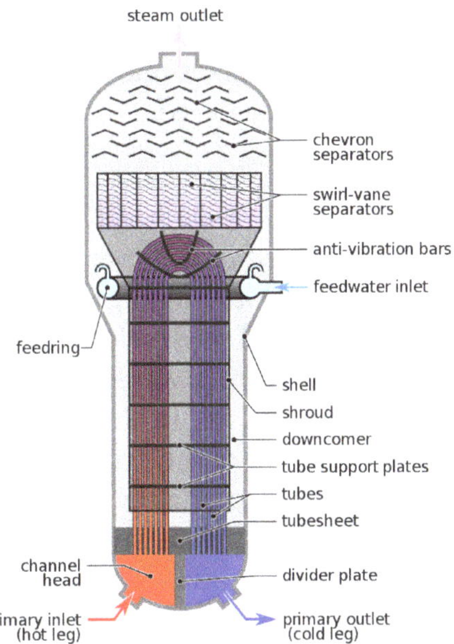

Vertical recirculating-type steam generator (typical of Westinghouse and Combustion Engineering-designed reactors) and components.

In commercial power plants, there are two to four steam generators per reactor; each steam generator can measure up to 70 feet (21 m) in height and weigh as much as 800 tons. Each steam generator can contain anywhere from 3,000 to 16,000 tubes, each about .75 inches (19 mm) in diameter. The coolant (treated water), which is maintained at high pressure to prevent boiling, is pumped through the nuclear reactor core. Heat transfer takes place between the reactor core and the circulating water and the coolant is then pumped through the primary tube side of the steam generator by coolant pumps before returning to the reactor core. This is referred to as the primary loop.

That water flowing through the steam generator boils water on the shell side (which is kept at a lower pressure than the primary side) to produce steam. This is referred to as the secondary loop. The secondary-side steam is delivered to the turbines to make electricity. The steam is subsequently condensed via cooled water from a tertiary loop and returned to the steam generator to be heated once again. The tertiary cooling water may be recirculated to cooling towers where it sheds waste heat before returning to condense more steam. Once-through tertiary cooling may otherwise be provided by a river, lake, or ocean. This primary, secondary, tertiary cooling scheme is the basis of the pressurized water reactor, which is the most common way to extract usable energy from a controlled nuclear reaction.

These loops also have an important safety role because they constitute one of the primary barriers between the radioactive and non-radioactive sides of the plant as the primary coolant becomes radioactive from its exposure to the core. For this reason, the integrity of the tubing is essential in minimizing the leakage of water between the primary and secondary sides of the plant. Steam generator tubes often degrade over time. There is the potential that, if a tube bursts while a plant is operating, contaminated steam could escape directly to the secondary cooling loop. Thus during scheduled maintenance outages or shutdowns, some or all of the steam generator tubes are inspected by eddy-current testing, and individual tubes can be plugged to remove them from operation.

Entire steam generators are often replaced in plant mid-life, which is a major undertaking. Most U.S. PWR plants have had steam generators replaced.

History

The nuclear powered steam generator started as a power plant for the first nuclear submarine, the USS *Nautilus* (SSN-571). It was designed and built by the Westinghouse power company for the submarine from there the company started its development and research of nuclear powered steam generators. Once peaceful nuclear reactors were legalized for use as power plants, power corporations jumped at the opportunity to utilize the growing development of nuclear powered steam generators. Westinghouse built one of the first nuclear power plants, the Yankee Rowe nuclear power station (NPS), which also used a nuclear powered steam generator, in 1960. This power plant had a one hundred MWe (mega watt electric) output. By comparison, some modern plants have over 1100 MWe output. Eventually, other international companies such as Babcock & Wilcox and Combustion Engineering began their own programs for research and development of the nuclear power steam generator. Since the 1960s, the US has fallen behind on some European nations in embracing this new power source. France and the UK have been more actively pursuing the benefits that come with nuclear energy, while the US is more concerned about the risk. Finally, it seems China is planning a massive increase to their nuclear power supply and ordering many new plants to be built.

Three Mile Island

In the Three Mile Island disaster, a main feed water pump shut down, although the cause is not known. Without that pump, the steam generator wasn't able to remove heat from the reactor, so pressure in the reactor began to rise. The system automatically began to dump water from the reactor to reduce pressure, but the relief valve got stuck open when the automation told it to close. The control room indicated that the valve was closed. The staff, therefore, had no idea that they were dumping radioactive water out of one of their reactors. With the water being pumped out, there wasn't enough emergency cooling water and the staff was unaware that they were losing more by the minute. Without adequate cooling, one of the reactors began to melt. The pipes burst and about half the core melted during the accident. Unlike Chernobyl and other nuclear disasters, the containment house around the reactor held and the damage to the outside world was very minimal. Since the containment housing held, no radioactive particles were released into the atmosphere, like what happened in Fukushima. The release of radioactive water did damage and contaminate the local area, but it did not spread from there.

Types

The giant vessel moves in a special train with speed restricted to 20 miles per hour.

This Babcock & Wilcox nuclear steam generator moved in a special train (restricted to 20 mph) via the Penn Central Railroad and Southern Railway from Barberton, Ohio to a Duke Energy site in Oconee, S.C. This generator weights 1,140,000 lbs and is a record shipment for the PC at that time (1970).

Westinghouse and Combustion Engineering designs have vertical U-tubes with inverted tubes for the primary water. Canadian, Japanese, French, and German PWR suppliers use the vertical configuration as well. Russian VVER reactor designs use horizontal steam generators, which have the tubes mounted horizontally. Babcock & Wilcox plants (e.g., Three Mile Island) have smaller steam generators that force water through the top of the OTSGs (once-through steam generators; counter-flow to the feedwater) and out the bottom to be recirculated by the reactor coolant pumps. The horizontal design has proven to be less susceptible to degradation than the vertical U-tube design.

Materials and Construction

The materials that make up the turbine and pipes of a nuclear powered steam generator are specially made and specifically designed to withstand the heat and radiation of the reactor. The water tubes also have to be able to resist corrosion from water for an extended period of time. The pipes that are used in American reactors are made of Inconel, either Alloy 600 or Alloy 690. Alloy 690 is made with extra chromium and most facilities heat treat the metal to make it better able to resist heat and corrosion. The high nickel content in Alloy 600 and Alloy 690 make them well suited for resisting acids and high degrees of stress and temperature.

Degradation

The annealed, or heat treated, Alloy 600 was prone to tube denting and thinning due to water chemistry. Plants that used the Alloy 600 in their water tubes therefore had to install new water chemistry controllers and change the chemicals they put in the water. Due to this, pipe thinning has been taken care of, but on rare occasions, tube denting still occurs, causing leaks and ruptures. The only way to prevent this is regular maintenance and check-ups, but this forces the reactor to shut down. In some cases, plants replaced their Alloy 600 tubes with Alloy 690 tubes and a few plants were shut down. To prevent future problems, manufacturers of steam turbines for nuclear power plants have improved their fabrication techniques and used other materials, such as stainless steel, to prevent tube denting.

Safety Devices

As the disaster at Three Mile Island could have been avoided with prior planning and a water level indicator, the Nuclear Regulatory Commission has started to push for water level controllers. The controller would regulate water to the reactor using a combination of sensors such as feedback controllers and feed-forward controllers. Yet this is only one of many such devices that ensure the safe and efficient production of power by nuclear reaction. Other tools include the control rods, relief valves, and even back up cooling systems. The control rods work by reducing the amount of radiation produced. They are built of a material that absorbs neutrons, and therefore reduces the number of fission reactions that take place inside the reactor. Relief valves function to vent pressure, sometimes into the atmosphere, in order to protect the system as a whole. And if the worst were to happen and a meltdown was occurring, a reservoir of cooling water waiting in standby could mean the difference between a disaster and a minor incident.

Typical Operating Conditions

Steam generators in a "typical" PWR in the USA have the following operating conditions:

Side	Pressure (absolute)	Inlet temperature	Outlet temperature
Primary side (tube side)	15.5 MPa (2,250 psi)	315 °C (599 °F) (liquid water)	275 °C (527 °F) (liquid water)
Secondary side (shell side)	6.2 MPa (900 psi)	220 °C (428 °F) (liquid water)	275 °C (527 °F) (saturated steam)

Tube Material

Various high-performance alloys and superalloys have been used for steam generator tubing, including type 316 stainless steel, Alloy 400, Alloy 600MA (mill annealed), Alloy 600TT (thermally treated), Alloy 690TT, and Alloy 800Mod.

Heat Transport

Shutdown of a nuclear reactor refers to termination of fission chain reaction. A nuclear reactor may be shutdown for one of the following reasons:

(i) The reactor is considered to have served for a time period for which it was designed

(ii) The reactor requires repairs

(iii) The reactor has suffered from a severe operational transient and must be prevented from suffering a severe accident.

A reactor is shutdown by inserting the control rods completely into the core or by introducing poisons like boron or both. Despite the termination of fission chain reaction, a considerable amount of heat is still present in the fuel rods in the form of sensible heat. Apart from this, heat is also generated due to decay of fission products and also due to fissions induced by delayed neutrons. Hence the nuclear reactor must be cooled for several hours even after shut down.

Let us look at the quantity of heat released after shutdown as a function of time after shutdown.

Fissions induced by delayed neutrons:

The heat released due to fissions induced by delayed neutrons decreases exponentially with time. In about 80-100 seconds, the heat released as a percentage of reactor power deceases to less than 2 %. Hence this constitutes only a small portion of shutdown cooling and can be removed in a short period of time.

Decay of fission products:

The heat released due to decay of fission products (also called decay heat) is prominent over a longer time period and hence determines the cooling requirements. The decay heat depends upon a number of factors including the number of days in which the reactor was in operation before shutdown, type of reactor etc.

This is approximately 7 % of reactor power (immediately after shutdown), 4 % after 100 seconds, 1.3 % after one hour, 0.5 % after 28 hours and 0.2 % after a week.

The normal heat removal system, if functioning properly, is sufficient for decay heat removal with a small change in the flow path. Under normal reactor operation, steam passes through the turbine and then to condenser before being sent back to core. However during decay heat removal, steam is directly sent to condenser bypassing the turbine. The condensate leaving the condenser is supplied back to the core, in case of BWR and to steam generator in case of PWR.

Shutdown Cooling System in BWR

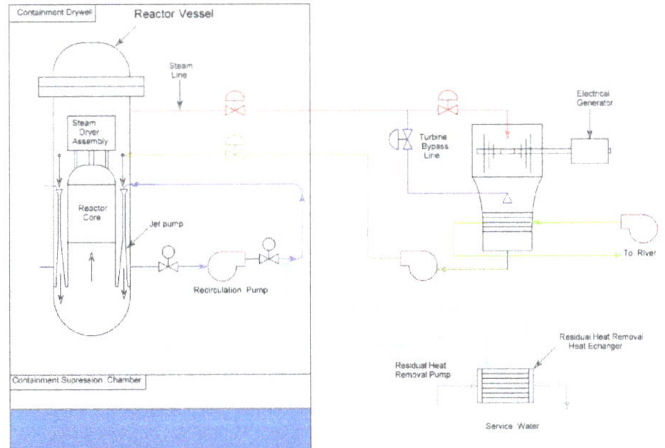

Schematic layout of shutdown cooling system of a typical BWR. The Residual Heat Removal System is also shown (Redrawn from Ref.)

A schematic diagram of shutdown cooling system in a typical Boiling Water Reactor is shown in Figure. In some boiling water reactors (similar to the one at Tarapur), an emergency condenser is provided which comes to operation in cases where the main condenser is isolated from the reactor. The emergency condenser has a huge pool of water that can act as heat sink to remove decay heat in a passive manner (by natural convection) for a period of about 8 hours. BWRs house a Residual heat removal system in which the hot liquid from core is pumped through an external heat exchanger and then returned back to core. This is utilized when the decay heat is insufficient for steam generation.

Shutdown Cooling System in PWR

A schematic diagram of shutdown cooling system in a typical Pressurized Water Reactor is shown in Figure.

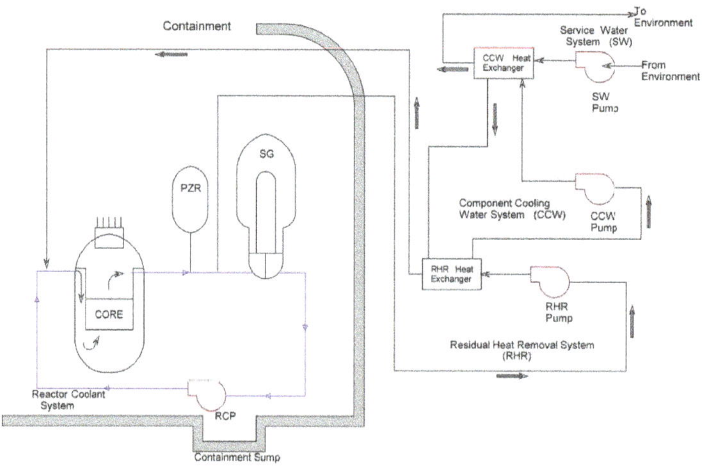

Schematic layout of shutdown cooling system of a typical PWR. The Residual Heat Removal System is also shown (Redrawn from Ref.)

In PWR, during decay heat removal, there are provisions for steam to be vented to atmosphere through atmospheric relief valves, in cases where cooling water for condenser is unavailable. When the decay heat is too low to generate sufficient steam, the Residual Heat Removal (RHR) system is used. In this system, a part of the primary coolant leaving the core is circulated through a heat exchanger to cool the primary coolant to the inlet temperature. This system brings the temperature to a very low value enabling personnel access.

Emergency Core Cooling System

Emergency core cooling system refers to the cooling system adopted in the cases of emergency. This could arise to substantial break in the primary coolant circuit that leads to the decrease in coolant pressure. ECCS becomes operative when the pressure in the primary coolant loop falls below the desired value.

ECCS contains several systems that function independent of each other. Each of these systems is designed to be activated at different pressure levels. All these systems introduce water that is supplied from different sources. Considerable amount of redundancy exists in these systems to ensure the reliability of ECCS. The arrangement of ECCS is different for different reactor types, taking into account of the core thermal-hydraulic characteristics and reactor physics aspects.

Let us discuss ECCS in pressurized water reactors, before we proceed with discussion on ECCS in boiling water reactors.

ECCS in Pressurized Water Reactors

The primary role of ECCS in PWR is of two fold: (i) To provide cooling to the core in order to minimize damage to the fuel, in the case of an accident caused by the loss of coolant. For this purpose, large amounts of low temperature, borated water is injected into the reactor (ii) To ensure that the reactor remains shutdown after the reactor has been cooled following a rupture of steam line. This is done by pumping water containing excess of neutron poisons stored in Refueling Water Storage Tank (RWST).

The Emergency Core Cooling System of a Pressurized water reactor consists of three subsystems. They are (i) High presssure and intermediate pressure injection system; (ii) accumulator injection system; (iii) low-pressure injection system. Each subsystem gets activated depending on the specific signals from the reactor say, coolant pressure. Each subsystem is provided with 2 pumps of equal capacity and capable of delivering the required flow individually. This is done to maintain redundancy. These pumps are operated using diesel generators and hence can operate even if the normal source of electricity for reactor is unavailable.

High Pressure and Intermediate Injection Systems

High-pressure injection system comes into operation in the case of moderate drop in pressure say from 155 bar to 110 bar. There are two conditions that lead to such a situation:

(a) small break in primary coolant circuit: This will result in reduced coolant flow rate. If un-noticed, this will result in appreciable boiling of coolant in the core which is undesirable in pressurized water reactors.

(b) failure of pressure relief valve in the pressurizer to close after being activated in an over-pressurization situation: Such a scenario may result in the loss of primary coolant, resulting in the boiling of coolant in core. It is relevant to talk about the "Three-Mile Island" accident at this juncture. In the reactor – 2 of the Three Mile Island plant, the pilot-operated relief valve was opened due to increase in steam pressure caused by reduced primary coolant flow. However, upon reaching the pressure close to normal operational limits, the valve must have closed. However, a mechanical fault led to non-closure of the relief valve. This valve in 'open position' allowed the escape of large quantities of water from the primary system, ultimately leading to fuel melt down.

Upon actuated, the high pressure injection system works by injecting borated water into reactor coolant inlet pipelines called cold legs, through the use of pumps driven by electricity. These pumps are the ones that are used in chemical and volume control system (CVCS). The acutation of this subsystem makes the pumps take input from refueling water storage tank (RWST) instead of from CVCS. Intermediate pressure system is activated in cases of moderate to intermediate size breaks in the primary coolant loop. This system too takes water from RWST. Hence high pressure and intermediate injection systems are sometimes considered together as a single subsystem in some cases.

Accumulator Injection System

In the event of a large break in primary coolant circuit, system pressure decreases rapidly due to which another ECCS subsystem called 'Accumulator injection system' gets actuated. The system contains large tanks called 'Accumulators' that contain cool borated water under nitrogen gas at a pressure of about 14 to 40 bar. These tanks are connected to cold leg (primary coolant inlet) via a non-return valve that gets actuated without manual intervention when the pressure in the cold leg is below that in accumulators. Hence this is a "passive system" that depends on gravity for discharge of borated water into the cold leg without the use of pumps.

Low-pressure Injection System

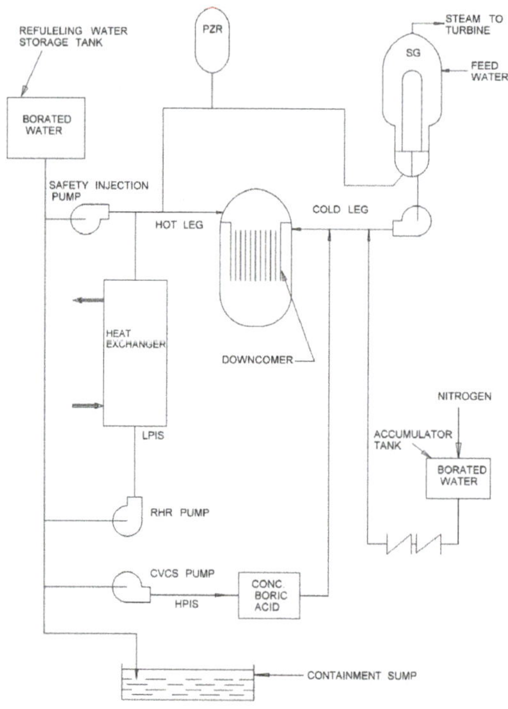

Schematic diagram of ECCS of a typical PWR (Redrawn from Ref.)

The Low-pressure injection system is also called the Residual Heat Removal system. In case of large breaks in primary coolant, the coolant pressure will go down to lower values acutating this subsystem. The pumps in this subsystem are also electrically operated and hence require AC power or power from diesel generators. Water is normally drawn from refuelling storage water tank. In case the water in RWST is exhasuted, required water may also be drawn from a sump located in the containment structure. Such water after passing through the core and extracting the heat, is cooled using a residual heat removal system heat exchanger and returns back to cool the reactor.

ECCS in Boiling Water Reactors

ECCS in BWR consists of three independent sub-systems. They are:

High Pressure Coolant Injection Systems

These inject large amount of water into the core to ensure that the core is covered with water. This system can inject water to the core even at high pressures. Under normal circumstances, core is always covered with water. However, breakage of coolant pipes may lead to loss of coolant, which if allowed to proceed without intervention would result in the exposure of core. In case of water level falling below safe limit, other sub-systems come to fore.

With the advancement in BWR technology, minor variations in this ECCS subsystem have been made. For example, BWR-6 manufactured by GE, uses a High-Pressure Core Spray, located between steam separator and top portion of the core, to spray water on to the core.

Automatic Depressurization System (ADS)

As the name indicates, this system is used to de-pressurize the system. This gets activated in the cases where HPCI alone is not capable of maintaining the requisite coolant level in the core while the reactor pressure is high. To ensure that the coolant level in maintained in the core, supply of coolant from sources other than HPCI must be facilitated. This is done by ADS which reduces pressure by releasing steam to a suppression pool containing water through pipes that connect the steam region with the suppression pool, where the steam is condensed. This brings the reactor pressure down to ~32 atm at which another ECCS subsystem, Low Pressure Coolant Injection (LPCI) system gets activated to supply coolant.

Low Pressure Coolant Injection (LPCI)

This subsystem injects large volumes of water when the reactor pressure falls below 32 atm, through the use of four diesel-driven pumps. This system is capable of flooding the core with water in a short period of time.

This subsystem has several variants depending upon the model of BWR. For example, Advanced Boiling Water Reactor (ABWR) has Low Pressure Core Flooder (LPCF) which functions on similar lines to that of LPCI.

Moderator and Moderator System

We have seen the use of various materials as moderators in different nuclear reactors. One may also recall a reactor classification based on the moderator. That leads us to two important questions: (i) what type of materials is suitable as moderator? (ii) What is the method of comparing the performance of different materials used as moderators?

To answer the first question, we shall recollect different type of nuclear reactions.

During elastic scattering, the speed and direction of a neutron are changed due to elastic collisions with a nucleus. The neutron looses a part of its kinetic energy and hence slows down. These are the thermal neutrons that help the nuclear chain reaction to sustain. This process is called moderation.

The nucleus with which the neutrons undergo elastic collision must be conducive for thermalization. In other words, the probability of elastic reaction between neutron and the nucleus must be high. In terms of cross sections, the nuclear cross section of a moderator's nucleus for elastic scattering (σ_{el}) must be high.

Using the analogy of collisions of billiard balls for elastic scattering, it is possible to relate the mass of target or moderator nucleus (M), energy of incident neutron (E_i) and the energy of scattered neutron (E_s) using the laws of conservation of mass and energy, as follows:

$$E_s = \left(\frac{M - m_n}{M + m_n}\right)^2 E_i$$

If the atomic mass of the target or moderator nucleus is approximated to its mass number (A) and mass of a neutron as 1 amu, then

$$E_s = \left(\frac{A-1}{A+1}\right)^2 E_i$$

Note: The simultaneous solution of momentum and energy conservation equations for elastic collision between two bodies (as in the case of collision of a neutron with a nucleus) gives two solutions:

Solution – I: Initial and final energies of neutron are same. In this case, there is no change in the direction as neutron continues to travel in forward direction. Hence $E_s=E_i$

Solution – II: The neutron is scattered at 180° i.e. the direction of neutron is completely reversed. The energies of incident and scattered neutrons are related by above equation

In reality, the direction of scattering may range from 0 to 180 °. Hence, the average energy of scattered neutron may be taken as the average of energies with scattering angle 0 and 180°.

Therefore,

$$E_s = \frac{E_i + \left(\frac{A-1}{A+1}\right)^2 E_i}{2} = E_i \frac{1 + \left(\frac{A-1}{A+1}\right)^2}{2}$$

$$\frac{E_s}{E_i} = \frac{1 + \left(\frac{A-1}{A+1}\right)^2}{2} = \frac{A^2 + 1}{A^2 + 2A + 1}$$

The above equation can be graphically represented as shown in figure.

From figure it is clear that lower the value of 'A', lower is the E_s/E_i ratio. Hence light nuclei are more effective in moderating compared to that of heavy nuclei. This is the rationale for using water (hydrogen as scattering nucleus), heavy water (deuterium as scattering nucleus) and graphite (carbon as scattering nucleus).

Above equations give the ratio of kinetic energies of scattered and incident neutron after one collision. However to thermalize the neutrons by reducing its energy from few MeV to 0.0253 eV (8-orders decrease in kinetic energy) the neutron must undergo several successive elastic collisions with the moderator.

The mean logarithmic energy decrement (ξ) defined as $\ln(E_i/E_s)$ is related to the mass number of moderator as:

$$\xi = 1 - \frac{(A-1)^2}{2A}\ln\left(\frac{A+1}{A-1}\right)$$

Higher the value of ξ, more effective the moderation is due to the lower energy of scattered neutron compared to the incident neutron. Equation (5) can be represented diagrammatically in Figure. It is clear from figure that ξ is high for nuclei of lower atomic number.

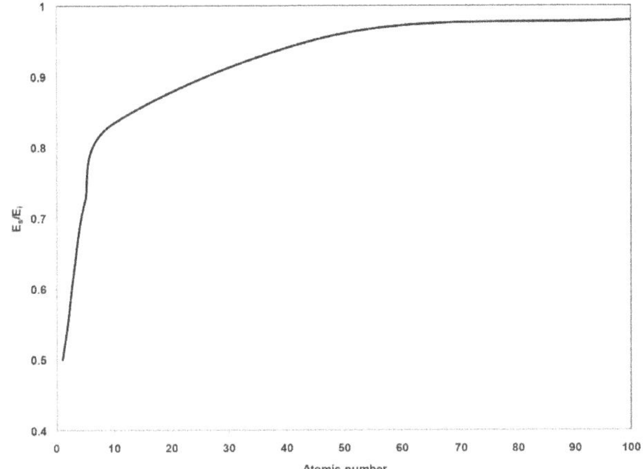

Ratio of energies of scattered neutrons to incident neutrons

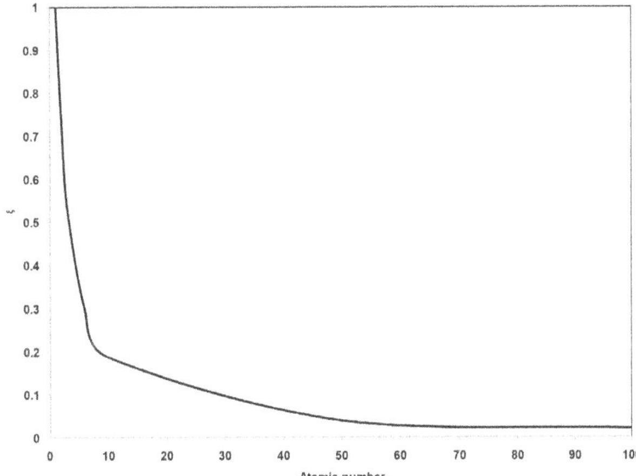

Mean logarithmic energy decrement as a function of atomic number

If E_{high} and $E_{thermal}$ are the kinetic energies of high energy neutron and thermal neutron respectively, then

$$\frac{E_{high}}{E_{thermal}} = \frac{E_{high}}{E_1}\cdot\frac{E_1}{E_2}\cdot\frac{E_2}{E_3}\cdots\cdots\frac{E_{n-1}}{E_{thermal}}$$

$$\frac{E_{high}}{E_{thermal}} = e^{\xi}.e^{\xi}.e^{\xi}...e^{\xi} = e^{n\xi}$$

The following example shows the calculation of nuclear of collisions for a typical case.

Example 1: Determine the number of collisions required for thermalization for the following case.

Data: Energy of incident neutron = E_{high} = 1 MeV

Energy of thermal neutron = $E_{thermal}$ = 0.0253 MeV

Moderator: Heavy water

$\xi = 0.509$

Solution:

Using Eq. (7)

$$\frac{1e6}{0.0253} = e^{0.509n}$$

Solving above Eq., we get n =34

For light water ($\xi = 0.920$) and graphite ($\xi = 0.158$) as moderators, the numbers of collisions required for reduction of neutron energy from 1 MeV to 0.0253 eV are 19 and 111 respectively (calculated using earlier equation).

Moderating ratio

It may be recalled that the nuclear cross sections represent the probability of various nuclear reactions. When neutron interacts with a nucleus, one of the following can happen: elastic scattering, inelastic scattering, neutron capture and fission.

As we have seen earlier, the slowing down of neutrons is due to the elastic scattering of neutron by the nucleus of moderator. If the capture cross section of moderator (σ_{γ}) is high, most of the neutrons will be absorbed by it, leading to lower moderation or lower availability of thermal neutrons.

Hence a higher ratio of scattering to capture cross sections ($\sigma_{el} / \sigma_{\gamma}$) is desirable for effective moderation. Combining this ratio with mean logarithmic energy decrement (ξ), 'moderating ratio' can be calculated which can be used as a criterion for comparison of different moderators.

$$Moderating\ ratio = \frac{\zeta \sigma_{el}}{\sigma_{\gamma}}$$

Moderating ratio will be high if either of (ξ) or ($\sigma_{e}l / \sigma_{\gamma}$) is higher. The following table gives the values of ξ, and $\sigma_{e}l / \sigma_{\gamma}$ of some common moderators.

Table: Moderating ratio for common moderators. Note that the cross sections are corresponding to the neutron energy of 0.0253 eV (thermal neutron).

Moderator	ξ	σ_{el} (b)	σ_{γ} (b)	$\sigma_{el}/\sigma_{\gamma}$	Moderating ratio	n
Light water	0.920	25.47	0.33	77.17	71	19
Heavy water	0.509	5.57	0.0005	11139.49	5670	34
Graphite	0.128	5.25	0.0035	1500	192	111

Though light water has the highest ξ and σ_{el} among the three moderators shown, its moderating ratio is low due to its relatively higher capture cross section. Heavy water has the highest moderating ratio owing to its lowest cross section for neutron capture. Graphite, despite possessing the lowest ξ and σ_{el}, fairs better than light water due to its lower σ compared to that of light water.

The relative advantages and disadvantages of heavy water and light water, the common moderators in Indian reactors are as follows:

Table: Relative advantages and disadvantages of light and heavy water as moderator.

Moderator	Advantages	Disadvantages
Heavy water	High moderating ratio and hence can be used with natural uranium fuel	More expensive; Requires a core of larger diameter to facilitate 34 collisions between a neutron and the moderator
Light water	Less expensive; Requires a core of relatively smaller diameter as only 19 collisions are required for moderation	Can be used only with enriched uranium fuel

Moderator System in Pressurized Heavy Water Reactor

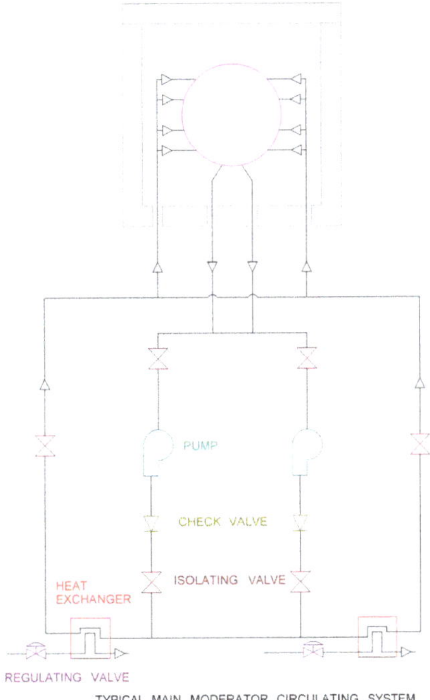

Moderator system of PHWR (Redrawn from Ref.)

Coolant and moderator are not separated in light water reactors. However, despite using heavy water as both moderator and coolant, they are housed in separate compartments in a pressurized heavy water reactor. Heavy water as coolant flows inside the pressure tubes, while the calandria is immersed in the moderator. The moderator is not directly heated by heat transfer from fuel. Instead, moderator is heated due to the energy lost by neutrons during their slow down. The energy of prompt gamma radiation is also deposited in the moderator. It may be recalled that about 5 % of 200 MeV of recoverable energy from fission is released in the moderator. Hence the moderator needs to be cooled to maintain its temperature, to prevent vaporization. Typically, the moderator is maintained between 60 °C and 80°C. The moderator system is designed to maintain the temperature of moderator within this range. A schematic diagram of moderator system with pumps, valves and heat exchangers is shown in Figure.

Calandria has outlets at the bottom for the removal of moderator. The inlets for moderator are provided on the sides of calandria as shown in Figure. Moderator pumps are used to pump the moderator out from calandria and pass them through heat exchangers. Moderator is cooled in these heat exchangers using light water supplied for this purpose. The flow rate of light water is controlled depending upon the temperature of moderator.

Two heat exchangers are shown in Figure. Each heat exchanger is provided with an isolation valve that can be used to disconnect that particular heat exchanger from the system, for maintenance. In such cases, one heat exchanger can be used to maintain the temperature of moderator. The presence of check valves in the line ensure that there is no return flow, in case of failure of moderator pump.

7

Energy: Sources and Spectrum

In fast reactors, the neutron energy is released in a wide spectrum. Unmoderated fast neutrons are used in these reactors since they have a better fission/capture ratio, making fast reactors breed more fissile fuel than it consumes. Nuclear science and technology is best understood in confluence with the major topics listed in the following chapter.

FBR Neutronics

One of the essential studies in neutronics of a fast reactor is the study of neutron energy spectrum. The neutron capture and fission occurs over a wide range of neutron energies in a fast reactor compared to that in a thermal reactor.

Lethargy (u) is defined as the natural logarithm of ratio of maximum energy that a neutron might have in a nuclear reactor (E_0) to the neutron energy (E). Hence, lethargy is a measure of neutron energy. The maximum neutron energy (E_0) is taken as 10 MeV.

$$Lethargy = u = ln\left(\frac{E_0}{E}\right) \quad (1)$$

Differentiating Eq. (1), we may relate the change in energy and change in lethargy as:

$$du = \frac{-dE}{E} \quad (2)$$

However, small changes in lethargy (Δu) may be written as

$$\Delta u = u_1 - u_2 = ln\left(\frac{E_2}{E_1}\right) \quad (3)$$

The neutron flux Vs neutron energy for typical fast reactor with metal oxide, metal carbide and metal as fuels is shown in Figure. The Y-axis is relative flux per unit lethargy. One may observe from Figure that the relative flux increases with neutron energy till about 3 keV. The flux decreases slightly around 3 keV, which is attributed to the higher scattering by sodium at this neutron energy level. After this small depression, the relative flux increases with neutron energy and reaches maximum around 100 keV for the oxide fuel. Decrease in relative flux is rapid beyond the neutron energies of 2 MeV.

One may observe from Figure, that at lower neutron energies (< 20-50 keV), the relative neutron flux per unit lethargy is highest for oxide fuel, followed by that for carbide and metal fuels. Above 100 keV, the metallic fuel has the highest relative flux among these fuels.

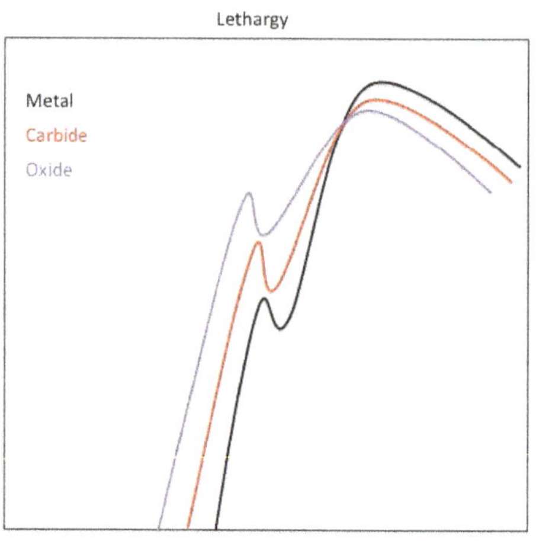

Relative neutron flux as a function of neutron energy

The neutron flux as a function of neutron energy is given by:

$$\phi(E) = C\sqrt{\frac{E}{(kT_m)^3}}\exp\left(\frac{-E}{kT_m}\right) \tag{4}$$

'kT_m' is taken as 1.4 MeV for Pu-239. 'kT_m' changes with the isotope.

The variation of neutron flux with neutron energy influences the fraction of fission that occurs at different energy levels. By dividing neutron energies in to different groups with known energy range for each group, it is possible to calculate the fraction of fission that can occur over different neutron energy ranges. The results of such a study for an oxide-fueled, 1200 MWe fast reactor is available in the literature. Few highlights from that table are:

(i) the fraction of fission over a energy range of 1.35-3.7 MeV is the largest (~0.18)

(ii) the neutron energy range over which a large fraction of fission occurs is 15 keV to 3.7 MeV, with the cumulative fission fraction in this range being 0.73

(iii) the median fission energy is ~ 150 eV. Median fission energy is the neutron energy above which the cumulative fission fraction is 0.5

(iv) Nearly 20 % of fission occurs with neutron energies lower than 10 keV for the oxide-fueled reactor. The fission fraction in this energy range for metal–fuelled reactor is very low. The larger Doppler coefficient in oxide-fueled fast reactor is attributed to relatively higher fission fraction (compared to metal-fueled reactors) at neutron energies lower than 10 keV.

In a nuclear reactor, the neutron energies vary from MeV (fast neutron) to 0.025 eV (thermal neutron). The neutron energy spectrum can be divided into three regions: (i) low-energy region (< 1 eV); (ii) resonance region (1 eV $<E_n<$1 MeV) and (iii) continuum region (0.01$<E_n<$25 MeV).

The cross sections for various neutron reactions are dependent on neutron energy levels. Let us discuss the variation of cross section with neutron energy for few important isotopes: U-235, Pu-239 and U-238.

Cross Sections for U-235

In the low energy region, the total cross section and the fission cross section for U- 235 decrease with neutron energy, except over a small neutron energy range near 0.25-0.30 eV. For instance, at 0.025 eV, the fission cross section for U-235 is 580 b while at 1 eV the fission cross section falls to 65 b. This explains the reason for thermalization of fast neutron to energies around 0.025 eV in a thermal reactor.

Looking at the continuum region in which fast reactors operate, the fission cross section decreases with neutron energy. For U-235, the fission cross section at 0.01 MeV is about 3.3 b while the same at 0.5 MeV is about 1.2 b. Between 0.5 MeV and 3 MeV, the fission cross section is about 1.2 to 1.3 b. Beyond 6 MeV, the fission cross section increases and reaches about 3 b at about 20 MeV. Table shows the cross sections for various neutron reactions for U-235 for different energy intervals of relevance to fast reactor.

Table: Cross sections for U-235

Energy interval	Capture cross section (b)	Fission cross section (b)
> 2.2 MeV	0.04	1.23
0.82 – 2.2 MeV	0.09	1.24
300 – 820 keV	0.18	1.18
110 – 300 keV	0.32	1.40
40 – 110 keV	0.53	1.74
15 – 40 keV	0.79	2.16
0.75 – 15 keV	1.71	4.36

Cross Sections for Pu-239

It may recalled that the Pu-239 is not a naturally occurring fissile isotope; instead produced in nuclear reactors by neutron capture and transmutation in U-238. The fission and total cross sections decrease with neutron energy in the low-energy region, except in the energy range between 0.2-0.4 eV due to resonance. The fission cross section at 0.025 eV is about 752 b while the capture cross section is about 270 b, with the absorption and total cross sections being 1022 and 1028 b respectively. The total cross section falls to less than 10 b when the neutron energies reach 1 MeV. The average total cross section in the neutron energy range of 1 MeV to 4 MeV is 7.7 b, while the fission cross section in this energy range is around 1.8 b. The capture cross section decreases sharply with neutron energy in the region between 1 MeV to 4 MeV, with the values being 0.05 b (1 MeV) and 0.001 b (4 MeV).

Table shows the cross sections for various neutron reactions for Pu-239 for different energy intervals of relevance to fast reactor.

Table: Cross sections for Pu-239

Energy interval	Capture cross section (b)	Fission cross section (b)
> 2.2 MeV	0.01	1.85
0.82 – 2.2 MeV	0.03	1.82
300 – 820 keV	0.11	1.60
110 – 300 keV	0.20	1.51
40 – 110 keV	0.35	1.60
15 – 40 keV	0.59	1.67
0.75 – 15 keV	1.98	2.78

Cross Sections for U-238

In the thermal region (<1 eV), the cross sections (total and fission) of U-235 and Pu- 239 decrease with neutron energy as:

$$\sigma \, \alpha \frac{1}{\sqrt{E}} \tag{5}$$

The capture cross section for U-238 in the thermal region scales with neutron energy in accordance with Equation 5. The cross sections for various neutron reactions for U- 238 for different energy intervals of relevance to fast reactor is shown in Table.

Table: Cross sections for U-238

Energy interval	Capture cross section (b)	Fission cross section (b)
> 2.2 MeV	0.01	0.58
0.82 – 2.2 MeV	0.09	0.02
300 – 820 keV	0.11	-
110 – 300 keV	0.15	-
40 – 110 keV	0.26	-
15 – 40 keV	0.47	-
0.75 – 15 keV	0.84	-

From the above discussion, it is clear that the cross section and neutron flux vary with the neutron energies. Hence the average cross section for a neutron reaction (absorption or fission) must account for the variation of cross section and neutron flux with neutron energy, as shown in Equation (6).

$$\sigma = \frac{\int \sigma(E)\phi(E)dE}{\int \phi(E)dE} \tag{6}$$

The following example illustrates the use of Equation (6) for the calculation of average fission cross section for Pu-239.

Energy range MeV	dE MeV	E MeV	$\phi(E)$=C*E*exp(-E) $cm^{-2}s^{-1}$	σ_f(E) b	σ_f(E) ϕ(E)dE	ϕ(E)dE
0.82 – 2.2	1.38	0.82	0.361C*	1.82	0.9071C*	0.4984C*
0.3 – 0.820	0.52	0.30	0.222C*	1.60	0.1849C*	0.1155C*
0.110 – 0.300	0.19	0.11	0.099C*	1.51	0.0283C*	0.0187C*
0.040 – 0.110	0.07	0.04	0.038C*	1.60	0.0043C*	0.0027C*

In the above table, Eq. (4) is re-written as:

$$\phi(E) = C\sqrt{\frac{E}{(kT_m)^3}}\exp\left(\frac{-E}{kT_m}\right) = C^*\sqrt{E}\exp(-E) \tag{7}$$

where C* is a lumped parameter which is constant for a given isotope

$$C^* = \sqrt{\frac{C}{(kT_m)^3}}\exp\left(\frac{1}{kT_m}\right) \tag{8}$$

The average fission cross section is

$$\sigma = \frac{\int \sigma(E)\phi(E)dE}{\int \phi(E)dE} = \frac{\sum \sigma(E)\phi(E)dE}{\sum \phi(E)dE} = 1.77\ b$$

Note: We do not need the value of 'C' or 'kT_m' for calculation of average cross Section

Breeder Reactor

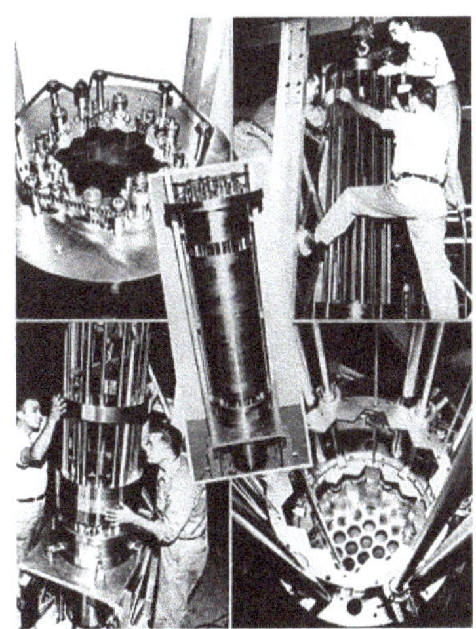

Assembly of the core of Experimental Breeder Reactor I in Idaho, United States, 1951

A breeder reactor is a nuclear reactor that generates more fissile material than it consumes. These devices achieve this because their neutron economy is high enough to breed more fissile fuel than they use from fertile material, such as uranium-238 or thorium-232. Breeders were at first found attractive because their fuel economy was better than light water reactors, but interest declined after the 1960s as more uranium reserves were found, and new methods of uranium enrichment reduced fuel costs.

Fuel Efficiency and Types of Nuclear Waste

Fission Probabilities of Selected Actinides, Thermal vs. Fast Neutrons				
Isotope	Thermal Fission Cross Section	Thermal Fission %	Fast Fission Cross Section	Fast Fission %
Th-232	nil	1 (non-fissile)	0.350 barn	3 (non-fissile)
U-232	76.66 barn	59	2.370 barn	95
U-233	531.2 barn	89	2.450 barn	93
U-235	584.4 barn	81	2.056 barn	80
U-238	11.77 microbarn	1 (non-fissile)	1.136 barn	11
Np-237	0.02249 barn	3 (non-fissile)	2.247 barn	27
Pu-238	17.89 barn	7	2.721 barn	70
Pu-239	747.4 barn	63	2.338 barn	85
Pu-240	58.77 barn	1 (non-fissile)	2.253 barn	55
Pu-241	1012 barn	75	2.298 barn	87
Pu-242	0.002557 barn	1 (non-fissile)	2.027 barn	53
Am-241	600.4 barn	1 (non-fissile)	0.2299 microbarn	21
Am-242m	6409 barn	75	2.550 barn	94
Am-243	0.1161 barn	1 (non-fissile)	2.140 barn	23
Cm-242	5.064 barn	1 (non-fissile)	2.907 barn	10
Cm-243	617.4 barn	78	2.500 barn	94
Cm-244	1.037 barn	4 (non-fissile)	0.08255 microbarn	33

Breeder reactors could, in principle, extract almost all of the energy contained in uranium or thorium, decreasing fuel requirements by a factor of 100 compared to widely used once-through light water reactors, which extract less than 1% of the energy in the uranium mined from the earth. The high fuel efficiency of breeder reactors could greatly reduce concerns about fuel supply or energy used in mining. Adherents claim that with seawater uranium extraction, there would be enough fuel for breeder reactors to satisfy our energy needs for 5 billion years at 1983's total energy consumption rate, thus making nuclear energy effectively a renewable energy.

Nuclear waste became a greater concern by the 1990s. In broad terms, spent nuclear fuel has two main components. The first consists of fission products, the leftover fragments of fuel atoms after they have been split to release energy. Fission products come in dozens of elements and hundreds of isotopes, all of them lighter than uranium. The second main component of spent fuel is transuranics (atoms heavier than uranium), which are generated from uranium or heavier atoms in the fuel when they absorb neutrons but do not undergo fission. All transuranic isotopes fall within the actinide series on the periodic table, and so they are frequently referred to as the actinides.

The physical behavior of the fission products is markedly different from that of the transuranics. In particular, fission products do not themselves undergo fission, and therefore cannot be used for nuclear weapons. Furthermore, only seven long-lived fission product isotopes have half-lives longer than a hundred years, which makes their geological storage or disposal less problematic than for transuranic materials.

With increased concerns about nuclear waste, breeding fuel cycles became interesting again because they can reduce actinide wastes, particularly plutonium and minor actinides. Breeder reactors are designed to fission the actinide wastes as fuel, and thus convert them to more fission products.

After "spent nuclear fuel" is removed from a light water reactor, it undergoes a complex decay profile as each nuclide decays at a different rate. Due to a physical oddity referenced below, there is a large gap in the decay half-lives of fission products compared to transuranic isotopes. If the transuranics are left in the spent fuel, after 1,000 to 100,000 years, the slow decay of these transuranics would generate most of the radioactivity in that spent fuel. Thus, removing the transuranics from the waste eliminates much of the long-term radioactivity of spent nuclear fuel.

Today's commercial light water reactors do breed some new fissile material, mostly in the form of plutonium. Because commercial reactors were never designed as breeders, they do not convert enough uranium-238 into plutonium to replace the uranium-235 consumed. Nonetheless, at least one-third of the power produced by commercial nuclear reactors comes from fission of plutonium generated within the fuel. Even with this level of plutonium consumption, light water reactors consume only part of the plutonium and minor actinides they produce, and nonfissile isotopes of plutonium build up, along with significant quantities of other minor actinides.

Conversion Ratio, Breakeven, Breeding Ratio, Doubling Time, and Burnup

One measure of a reactor's performance is the "conversion ratio" (the average number of new fissile atoms created per fission event). All proposed nuclear reactors except specially designed and operated actinide burners experience some degree of conversion. As long as there is any amount of a fertile material within the neutron flux of the reactor, some new fissile material is always created.

The ratio of new fissile material in spent fuel to fissile material consumed from the fresh fuel is known as the "conversion ratio" or "breeding ratio" of a reactor.

For example, commonly used light water reactors have a conversion ratio of approximately 0.6. Pressurized heavy water reactors (PHWR) running on natural uranium have a conversion ratio of 0.8. In a breeder reactor, the conversion ratio is higher than 1. "Breakeven" is achieved when the conversion ratio becomes 1: the reactor produces as much fissile material as it uses.

"Doubling time" is the amount of time it would take for a breeder reactor to produce enough new fissile material to create a starting fuel load for another nuclear reactor. This was considered an important measure of breeder performance in early years, when uranium was thought to be scarce. However, since uranium is more abundant than thought, and given the amount of plutonium available in spent reactor fuel, doubling time has become a less important metric in modern breeder reactor design.

"Burnup" is a measure of how much energy has been extracted from a given mass of heavy metal in fuel, often expressed (for power reactors) in terms of gigawatt-days per ton of heavy metal. Burnup is an important factor in determining the types and abundances of isotopes produced by a fission reactor. Breeder reactors, by design, have extremely high burnup compared to a conventional reactor, as breeder reactors produce much more of their waste in the form of fission products, while most or all of the actinides are meant to be fissioned and destroyed.

In the past breeder reactor development focused on reactors with low breeding ratios, from 1.01 for the Shippingport Reactor running on thorium fuel and cooled by conventional light water to over 1.2 for the Soviet BN-350 liquid-metal-cooled reactor. Theoretical models of breeders with liquid sodium coolant flowing through tubes inside fuel elements ("tube-in-shell" construction) suggest breeding ratios of at least 1.8 are possible on an industrial scale. The Soviet BR-1 test reactor achieved a breeding ratio of 2.5 under non-commercial conditions.

Types of Breeder Reactor

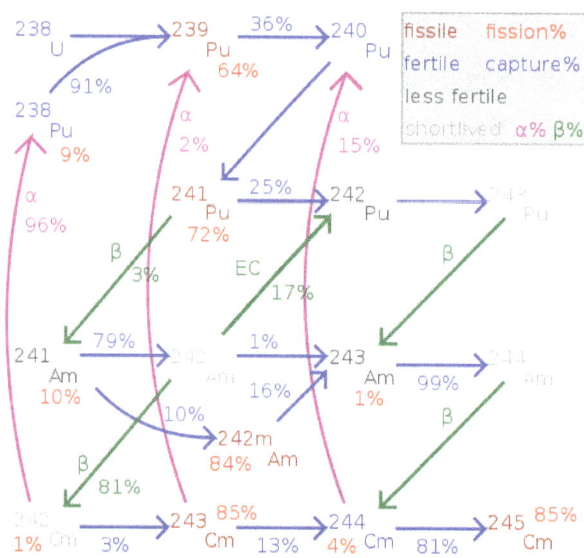

Production of heavy transuranic actinides in current thermal-neutron fission reactors through neutron capture and decays. Starting at uranium-238, isotopes of plutonium, americium, and curium are all produced. In a fast neutron breeder reactor, all these isotopes may be burned as fuel.

Many types of breeder reactor are possible:

A 'breeder' is simply a reactor designed for very high neutron economy with an associated conversion rate higher than 1.0. In principle, almost any reactor design could possibly be tweaked to become a breeder. An example of this process is the evolution of the Light Water Reactor, a very heavily moderated thermal design, into the Super Fast Reactor concept, using light water in an extremely low-density supercritical form to increase the neutron economy high enough to allow breeding.

Aside from water cooled, there are many other types of breeder reactor currently envisioned as possible. These include molten-salt cooled, gas cooled, and liquid metal cooled designs in many variations. Almost any of these basic design types may be fueled by uranium, plutonium, many minor actinides, or thorium, and they may be designed for many different goals, such as creating more fissile fuel, long-term steady-state operation, or active burning of nuclear wastes.

For convenience, it is perhaps simplest to divide the extant reactor designs into two broad categories based upon their neutron spectrum, which has the natural effect of dividing the reactor designs into those designed to use primarily uranium and transuranics, and those designed to use thorium and avoid transuranics.

- Fast breeder reactor or FBR uses fast (unmoderated) neutrons to breed fissile plutonium and possibly higher transuranics from fertile uranium-238. The fast spectrum is flexible enough that it can also breed fissile uranium-233 from thorium, if desired.
- Thermal breeder reactor use thermal spectrum (moderated) neutrons to breed fissile uranium-233 from thorium (thorium fuel cycle). Due to the behavior of the various nuclear fuels, a thermal breeder is thought commercially feasible only with thorium fuel, which avoids the buildup of the heavier transuranics.

Reprocessing

Fission of the nuclear fuel in any reactor produces neutron-absorbing fission products. Because of this unavoidable physical process, it is necessary to reprocess the fertile material from a breeder reactor to remove those neutron poisons. This step is required if one is to fully utilize the ability to breed as much or more fuel than is consumed. All reprocessing can present a proliferation concern, since it extracts weapons usable material from spent fuel. The most common reprocessing technique, PUREX, presents a particular concern, since it was expressly designed to separate pure plutonium. Early proposals for the breeder reactor fuel cycle posed an even greater proliferation concern because they would use PUREX to separate plutonium in a highly attractive isotopic form for use in nuclear weapons.

Several countries are developing reprocessing methods that do not separate the plutonium from the other actinides. For instance, the non-water based pyrometallurgical electrowinning process, when used to reprocess fuel from an integral fast reactor, leaves large amounts of radioactive actinides in the reactor fuel. More conventional advanced, water-based reprocessing systems like PUREX, include SANEX, UNEX, DIAMEX, COEX, and TRUEX—as well as proposals to combine PUREX with co-processes.

All these systems have modestly better proliferation resistance than PUREX, though their adoption rate is low.

In the thorium cycle, thorium-232 breeds by converting first to protactinium-233, which then decays to uranium-233. If the protactinium remains in the reactor, small amounts of U-232 are also produced, which has the strong gamma emitter Tl-208 in its decay chain. Similar to uranium-fueled designs, the longer the fuel and fertile material remain in the reactor, the more of these undesirable elements build up. Inside the envisioned commercial thorium reactors high levels of U232 would be allowed to accumulate, leading to extremely high gamma radiation doses from any uranium derived from thorium. These gamma rays complicate the safe handling of a weapon and the design of its electronics; this explains why U-233 has never been pursued for weapons beyond proof-of-concept demonstrations.

Waste Reduction

Actinides and fission products by half-life								
Actinides by decay chain				Half-life range (y)	Fission products of ^{235}U by yield			
4*n*	4*n*+1	4*n*+2	4*n*+3			4.5–7%	0.04–1.25%	<0.001%
^{228}Ra$^{№}$				4–6	†		^{155}Eu^{þ}	
^{244}Cm^{f}	^{241}Pu^{f}	^{250}Cf	^{227}Ac$^{№}$	10–29		^{90}Sr	^{85}Kr	^{113m}Cd^{þ}
^{232}U^{f}		^{238}Pu$^{f№}$	^{243}Cm^{f}	29–97		^{137}Cs	^{151}Sm^{þ}	^{121m}Sn
^{248}Bk	^{249}Cf^{f}	^{242m}Am^{f}		141–351	No fission products have a half-life in the range of 100–210 k years ...			
	^{241}Am^{f}		^{251}Cf^{f}	430–900				
		^{226}Ra$^{№}$	^{247}Bk	1.3 k – 1.6 k				
^{240}Pu$^{f№}$	^{229}Th$^{№}$	^{246}Cm^{f}	^{243}Am^{f}	4.7 k – 7.4 k				
	^{245}Cm^{f}	^{250}Cm		8.3 k – 8.5 k				
			^{239}Pu$^{f№}$	24.1 k				
		^{230}Th$^{№}$	^{231}Pa$^{№}$	32 k – 76 k				
^{236}Np^{f}	^{233}U$^{f№}$	^{234}U$^{№}$		150 k – 250 k	‡	^{99}Tc$^{¢}$	^{126}Sn	
^{248}Cm		^{242}Pu^{f}		327 k – 375 k			^{79}Se$^{¢}$	
				1.53 M		^{93}Zr		
	^{237}Np$^{f№}$			2.1 M – 6.5 M		^{135}Cs$^{¢}$	^{107}Pd	
^{236}U$^{№}$			^{247}Cm^{f}	15 M – 24 M			^{129}I$^{¢}$	
^{244}Pu$^{№}$				80 M	... nor beyond 15.7 M years			
^{232}Th$^{№}$		^{238}U$^{№}$	^{235}U$^{f№}$	0.7 G – 14.1 G				

Legend for superscript symbols
¢ has thermal neutron capture cross section in the range of 8–50 barns
f fissile
m metastable isomer
№ naturally occurring radioactive material (NORM)
þ neutron poison (thermal neutron capture cross section greater than 3k barns)
† range 4–97 y: Medium-lived fission product
‡ over 200,000 y: Long-lived fission product

Nuclear waste became a greater concern by the 1990s. Breeding fuel cycles attracted renewed interest because of their potential to reduce actinide wastes, particularly plutonium and minor actinides. Since breeder reactors on a closed fuel cycle would use nearly all of the actinides fed into them as fuel, their fuel requirements would be reduced by a factor of about 100. The volume of waste they generate would be reduced by a factor of about 100 as well. While there is a huge reduction in the *volume* of waste from a breeder reactor, the *activity* of the waste is about the same as that produced by a light water reactor.

In addition, the waste from a breeder reactor has a different decay behavior, because it is made up of different materials. Breeder reactor waste is mostly fission products, while light water reactor waste has a large quantity of transuranics. After spent nuclear fuel has been removed from a light water reactor for longer than 100,000 years, these transuranics would be the main source of radioactivity. Eliminating them would eliminate much of the long-term radioactivity from the spent fuel.

In principle, breeder fuel cycles can recycle and consume all actinides, leaving only fission products. Fission products have a peculiar 'gap' in their aggregate half-lives, such that no fission products have a half-life longer than 91 years and shorter than two hundred thousand years. As a result of this physical oddity, after several hundred years in storage, the activity of the radioactive waste from a Fast Breeder Reactor would quickly drop to the low level of the long-lived fission products. However, to obtain this benefit requires the highly efficient separation of transuranics from spent fuel. If the fuel reprocessing methods used leave a large fraction of the transuranics in the final waste stream, this advantage would be greatly reduced.

Both types of breeding cycles can reduce actinide wastes:

- The fast breeder reactor's fast neutrons can fission actinide nuclei with even numbers of both protons and neutrons. Such nuclei usually lack the low-speed "thermal neutron" resonances of fissile fuels used in LWRs.
- The thorium fuel cycle inherently produces lower levels of heavy actinides. The fertile material in the thorium fuel cycle has an atomic weight of 232, while the fertile material in the uranium fuel cycle has an atomic weight of 238. That mass difference means that thorium-232 requires six more neutron capture events per nucleus before the transuranic elements can be produced. In addition to this simple mass difference, the reactor gets two chances to fission the nuclei as the mass increases: First as the effective fuel nuclei U233, and as it absorbs two more neutrons, again as the fuel nuclei U235.

A reactor whose main purpose is to destroy actinides, rather than increasing fissile fuel stocks, is sometimes known as a burner reactor. Both breeding and burning depend on good neutron economy, and many designs can do either. Breeding designs surround the core by a breeding blanket of fertile material. Waste burners surround the core with non-fertile wastes to be destroyed. Some designs add neutron reflectors or absorbers.

Breeder Reactor Concepts

There are several concepts for breeder reactors; the two main ones are:

- Reactors with a fast neutron spectrum are called fast breeder reactors (FBR) – these typically utilize uranium-238 as fuel.
- Reactors with a thermal neutron spectrum are called thermal breeder reactors – these typically utilize thorium-232 as fuel.

Fast Breeder Reactor

In 2006 all large-scale fast breeder reactor (FBR) power stations were liquid metal fast breeder reactors (LMFBR) cooled by liquid sodium. These have been of one of two designs:

- *Loop* type, in which the primary coolant is circulated through primary heat exchangers outside the reactor tank (but inside the biological shield due to radioactive sodium-24 in the primary coolant).

Experimental Breeder Reactor II, which served as the prototype for the Integral Fast Reactor

- *Pool* type, in which the primary heat exchangers and pumps are immersed in the reactor tank.

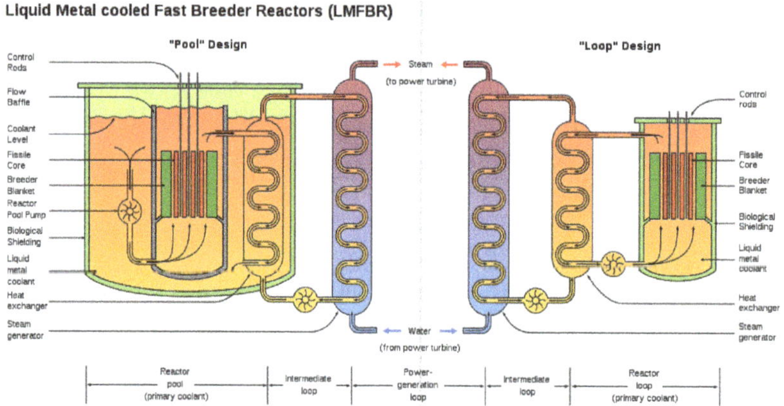

Schematic diagram showing the difference between the Loop and Pool types of LMFBR.

All current fast neutron reactor designs use liquid metal as the primary coolant, to transfer heat from the core to steam used to power the electricity generating turbines. FBRs have been built cooled by liquid metals other than sodium—some early FBRs used mercury, other experimental reactors have used a sodium-potassium alloy called NaK. Both have the advantage that they are liquids at room temperature, which is convenient for experimental rigs but less important for pilot or full scale power stations. Lead and lead-bismuth alloy have also been used. The relative merits of lead vs sodium are discussed here. Looking further ahead, four of the proposed generation IV reactor types are FBRs:

- Gas-Cooled Fast Reactor (GFR) cooled by helium.
- Sodium-Cooled Fast Reactor (SFR) based on the existing Liquid Metal FBR (LMFBR) and Integral Fast Reactor designs.
- Lead-Cooled Fast Reactor (LFR) based on Soviet naval propulsion units.
- Supercritical Water Reactor (SCWR) based on existing LWR and supercritical boiler technology.

FBRs usually use a mixed oxide fuel core of up to 20% plutonium dioxide (PuO_2) and at least 80% uranium dioxide (UO_2). Another fuel option is metal alloys, typically a blend of uranium, plutonium, and zirconium (used because it is "transparent" to neutrons). Enriched uranium can also be used on its own.

Many designs surround the core in a blanket of tubes that contain non-fissile uranium-238, which, by capturing fast neutrons from the reaction in the core, converts to fissile plutonium-239 (as is some of the uranium in the core), which is then reprocessed and used as nuclear fuel. Other FBR designs rely on the geometry of the fuel itself (which also contains uranium-238), arranged to attain sufficient fast neutron capture. The plutonium-239 (or the fissile uranium-235) fission cross-section is much smaller in a fast spectrum than in a thermal spectrum, as is the ratio between the $^{239}Pu/^{235}U$ fission cross-section and the ^{238}U absorption cross-section. This increases the concentration of $^{239}Pu/^{235}U$ needed to sustain a chain reaction, as well as the ratio of breeding to fission.

On the other hand, a fast reactor needs no moderator to slow down the neutrons at all, taking advantage of the fast neutrons producing a greater number of neutrons per fission than slow neutrons. For this reason ordinary liquid water, being a moderator as well as a neutron absorber, is an undesirable primary coolant for fast reactors. Because large amounts of water in the core are required to cool the reactor, the yield of neutrons and therefore breeding of ^{239}Pu are strongly affected. Theoretical work has been done on reduced moderation water reactors, which may have a sufficiently fast spectrum to provide a breeding ratio slightly over 1. This would likely result in an unacceptable power derating and high costs in an liquid-water-cooled reactor, but the supercritical water coolant of the SCWR has sufficient heat capacity to allow adequate cooling with less water, making a fast-spectrum water-cooled reactor a practical possibility.

The only commercially operating reactor as of 2015 is the BN-600 reactor in Russia, a 560MW sodium cooled reactor.

Integral Fast Reactor

One design of fast neutron reactor, specifically designed to address the waste disposal and plutonium issues, was the *integral fast reactor* (also known as an *integral fast breeder reactor*, although the original reactor was designed to not breed a net surplus of fissile material).

To solve the waste disposal problem, the IFR had an on-site electrowinning fuel reprocessing unit that recycled the uranium and all the transuranics (not just plutonium) via electroplating, leaving just short half-life fission products in the waste. Some of these fission products could later be separated for industrial or medical uses and the rest sent to a waste repository (where they would not have to be stored for anywhere near as long as wastes containing long half-life transuranics). The IFR pyroprocessing system uses molten cadmium cathodes and electrorefiners to reprocess metallic fuel directly on-site at the reactor. Such systems not only commingle all the minor actinides with both uranium and plutonium, they are compact and self-contained, so that no plutonium-containing material ever needs to be transported away from the site of the breeder reactor. Breeder reactors incorporating such technology would most likely be designed with breeding ratios very close to 1.00, so that after an initial loading of enriched uranium and/or plutonium fuel, the reactor would then be refueled only with small deliveries of natural uranium metal. A quantity of natural uranium metal equivalent to a block about the size of a milk crate delivered once per month would be all the fuel such a 1 gigawatt reactor would need. Such self-contained breeders are currently envisioned as the final self-contained and self-supporting ultimate goal of nuclear reactor designers. The project was canceled in 1994 by United States Secretary of Energy Hazel O'Leary.

Other Fast Reactors

The graphite core of the Molten Salt Reactor Experiment

Another proposed fast reactor is a fast molten salt reactor, in which the molten salt's moderating properties are insignificant. This is typically achieved by replacing the light metal fluorides (e.g. LiF, BeF_2) in the salt carrier with heavier metal chlorides (e.g., KCl, RbCl, $ZrCl_4$).

Several prototype FBRs have been built, ranging in electrical output from a few light bulbs' equivalent (EBR-I, 1951) to over 1,000 MWe. As of 2006, the technology is not economically competitive to thermal reactor technology—but India, Japan, China, South Korea and Russia are all committing substantial research funds to further development of Fast Breeder reactors, anticipating that rising uranium prices will change this in the long term. Germany, in contrast, abandoned the technology due to safety concerns. The SNR-300 fast breeder reactor was finished after 19 years despite cost overruns summing up to a total of 3.6 billion Euros, only to then be abandoned.

As well as their thermal breeder program, India is also developing FBR technology, using both uranium and thorium feedstocks.

Thermal Breeder Reactor

The advanced heavy water reactor (AHWR) is one of the few proposed large-scale uses of thorium. India is developing this technology, their interest motivated by substantial thorium reserves; almost a third of the world's thorium reserves are in India, which also lacks significant uranium reserves.

The third and final core of the Shippingport Atomic Power Station 60 MWe reactor was a light water thorium breeder, which began operating in 1977. It used pellets made of thorium dioxide and uranium-233 oxide; initially, the U-233 content of the pellets was 5–6% in the seed region, 1.5–3% in the blanket region and none in the reflector region. It operated at 236 MWt, generating 60 MWe and ultimately produced over 2.1 billion kilowatt hours of electricity. After five years, the core was removed and found to contain nearly 1.4% more fissile material than when it was installed, demonstrating that breeding from thorium had occurred.

The Shippingport Reactor, used as a prototype Light Water Breeder for five years beginning in August, 1977

The liquid fluoride thorium reactor (LFTR) is also planned as a thorium thermal breeder. Liquid-fluoride reactors may have attractive features, such as inherent safety, no need to manufacture fuel rods and possibly simpler reprocessing of the liquid fuel. This concept was first investigated at the Oak Ridge National Laboratory Molten-Salt Reactor Experiment in the 1960s. From 2012 it became the subject of renewed interest worldwide. Japan, China, the UK, as well as private US, Czech and Australian companies have expressed intent to develop and commercialize the technology.

Discussion

Like many aspects of nuclear power, fast breeder reactors have been subject to much controversy over the years. In 2010 the International Panel on Fissile Materials said "After six decades and the expenditure of the equivalent of tens of billions of dollars, the promise of breeder reactors remains largely unfulfilled and efforts to commercialize them have been steadily cut back in most countries". In Germany, the United Kingdom, and the United States, breeder reactor development programs have been abandoned. The rationale for pursuing breeder reactors—sometimes explicit and sometimes implicit—was based on the following key assumptions:

- It was expected that uranium would be scarce and high-grade deposits would quickly become depleted if fission power were deployed on a large scale; the reality, however, is that since the end of the cold war, uranium has been much cheaper and more abundant than early designers expected.
- It was expected that breeder reactors would quickly become economically competitive with the light-water reactors that dominate nuclear power today, but the reality is that capital costs are at least 25% more than water cooled reactors.
- It was thought that breeder reactors could be as safe and reliable as light-water reactors, but safety issues are cited as a concern with fast reactors that use a sodium coolant, where a leak could lead to a sodium fire.

- It was expected that the proliferation risks posed by breeders and their "closed" fuel cycle, in which plutonium would be recycled, could be managed. But since plutonium breeding reactors produce plutonium from U238, and thorium reactors produce fissile U233 from thorium, all breeding cycles could theoretically pose proliferation risks. U232, which is always present in U233 produced in breeder reactors, however is a strong alpha emitter, and as such would make weapon handling extremely hazardous, and the weapon easy to detect.

There are some past anti-nuclear advocates that have become pro-nuclear power as a clean source of electricity since breeder reactors effectively recycle most of their waste. This solves one of the most important negative issues of nuclear power. In the documentary *Pandora's Promise*, a case is made for breeder reactors because they provide a real, high kW alternative to fossil fuel energy. According to the movie, one pound of uranium provides as much energy as 5000 barrels of oil.

FBRs have been built and operated in the United States, the United Kingdom, France, the former USSR, India and Japan. The experimental FBR SNR-300 was built in Germany but never operated and eventually shut down amid political controversy following the Chernobyl disaster. As of 2014 one such reactor was being used for power generation, with another scheduled for early 2015. Several reactors are planned, many for research related to the Generation IV reactor initiative.

Development and Notable Breeder Reactors

Notable Breeder reactors											
Reactor	Country when built	Started	Shutdown	Design MWe	Final MWe	Thermal Power MWt	Capacity factor	No of leaks	Neutron temperature	Coolant	Reactor class
DFR	UK	1962	1977	14	11	65	34%	7	Fast	NaK	Test
BN-350	Soviet Union	1973	1999	135	52	750	43%	15	Fast	Sodium	Prototype
Rapsodie	France	1967	1983	0	–	40	–	2	Fast	Sodium	Test
Phénix	France	1975	2010	233	130	563	40.5%	31	Fast	Sodium	Prototype
PFR	UK	1976	1994	234	234	650	26.9%	20	Fast	Sodium	Prototype
KNK II	Germany	1977	1991	18	17	58	17.1%	21	Fast	Sodium	Research/Test
SNR-300	Germany	1985 (partial operation)	1991	327	–	–	–	–	Fast	Sodium	Prototype/Commercial
BN-600	Soviet Union	1981	operating	560	560	1470	74.2%	27	Fast	Sodium	Prototype/Commercial(Gen2)
FFTF	USA	1982	1993	0	–	400	–	1	Fast	Sodium	Test
Superphénix	France	1985	1998	1200	1200	3000	7.9%	7	Fast	Sodium	Prototype/Commercial(Gen2)
FBTR	India	1985	operating	13	–	40	–	6	Fast	Sodium	Test

PFBR	India	commissioning	commissioning	500	–	1250	–	–	Fast	Sodium	Prototype/ Commercial(Gen3)
Jōyō	Japan	1977	operating	0	–	150	–	–	Fast	Sodium	Test
Monju	Japan	1995	dormant	246	246	714	trial only	1	Fast	Sodium	Prototype
BN-800	Russia	2015	operating	789	880	2100	–	–	Fast	Sodium	Prototype/ Commercial(Gen3)
MSRE	USA	1965	1969	0	–	7.4	–	–	Epithermal	Molten Salt(FLiBe)	Test
Clementine	USA	1946	1952	0	–	0.025	–	–	Fast	Mercury	World's First Fast Reactor
EBR-1	USA	1951	1964	0.2	0.2	1.4	–	–	Fast	NaK	World's First Power Reactor
Fermi-1	USA	1963	1972	66	66	200	–	–	Fast	Sodium	Prototype
EBR-2	USA	1964	1994	19	19	62.5	–	–	Fast	Sodium	Experimental/ Test
Shippingport	USA	1977 as breeder	1982	60	60	236	–	–	Thermal	Light Water	Experimental-Core3

The Soviet Union (comprising Russia and other countries, dissolved in 1991) constructed a series of fast reactors, the first being mercury-cooled and fueled with plutonium metal, and the later plants sodium-cooled and fueled with plutonium oxide.

BR-1 (1955) was 100W (thermal) was followed by BR-2 at 100 kW and then the 5MW BR-5.

BOR-60 (first criticality 1969) was 60 MW, with construction started in 1965.

Future Plants

India has been an early leader in the FBR segment and is currently rolling out the next phase of their program. In 2012 an FBR called the Prototype Fast Breeder Reactor was under construction in India, due to be completed that year, with commissioning date known by mid-year. The FBR program of India includes the concept of using fertile thorium-232 to breed fissile uranium-233. India is also pursuing the thorium thermal breeder reactor. A thermal breeder is not possible with purely uranium/ plutonium based technology. Thorium fuel is the strategic direction of the power program of India, owing to the nation's large reserves of thorium, but worldwide known reserves of thorium are also some four times those of uranium. India's Department of Atomic Energy (DAE) said in 2007 that it would simultaneously construct four more breeder reactors of 500 MWe each including two at Kalpakkam.

BHAVINI, an Indian nuclear power company was established in 2003 responsible for the construction, commissioning and operation of all Stage II fast breeder reactors envisaged as part of the country's three stage nuclear power programme.

The new Indian FBR-600 is being used to advance these plans. The FBR-600 is a pool-type sodium cooled reactor with a rating of 600MWe and advanced active and passive safety features.

The Chinese Experimental Fast Reactor is a 65 MW (thermal), 20 MW (electric), sodium-cooled, pool-type reactor with a 30-year design lifetime and a target burnup of 100 MWd/kg.

The China Experimental Fast Reactor (CEFR) is a 25 MW(e) prototype for the planned China Prototype Fast Reactor (CFRP). It started generating power on 21 July 2011.

China also initiated a research and development project in thorium molten-salt thermal breeder reactor technology (Liquid fluoride thorium reactor), formally announced at the Chinese Academy of Sciences (CAS) annual conference in January 2011. Its ultimate target is to investigate and develop a thorium-based molten salt nuclear system over about 20 years.

Kirk Sorensen, former NASA scientist and Chief Nuclear Technologist at Teledyne Brown Engineering, has long been a promoter of thorium fuel cycle and particularly liquid fluoride thorium reactors. In 2011, Sorensen founded Flibe Energy, a company aimed to develop 20–50 MW LFTR reactor designs to power military bases.

South Korea is developing a design for a standardized modular FBR for export, to complement the standardized PWR (Pressurized Water Reactor) and CANDU designs they have already developed and built, but has not yet committed to building a prototype.

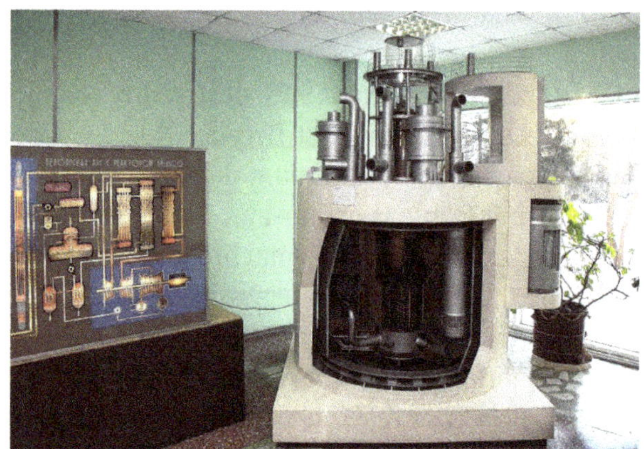

A cutaway model of the BN-600 reactor, superseded by the BN-800 reactor family.

Russia has a plan for increasing its fleet of fast breeder reactors significantly. A BN-800 reactor (800 MWe) at Beloyarsk was completed in 2012, succeeding a smaller BN-600. In June 2014 the

BN-800 was started in the minimum power mode. The reactor working at 35% of the nominal efficiency, was plugged-in the energy network on 10 December 2015.

Plans for the construction of an even larger BN-1200 reactor (1,200 MWe) initially anticipated completion in 2018, with two additional BN-1200 reactors built by the end of 2030. However, in 2015 Rosenergoatom postponed construction indefinitely to allow fuel design to be improved after more experience of operating the BN-800 reactor, and amongst cost concerns.

An experimental lead-cooled fast reactor, BREST-300 will be built at the Siberian Chemical Combine (SCC) in Seversk. The BREST design is seen as a successor to the BN series and the 300 MWe unit at the SCC could be the forerunner to a 1,200 MWe version for wide deployment as a commercial power generation unit. The development program is as part of an Advanced Nuclear Technologies Federal Program 2010-2020 that seeks to exploit fast reactors as a way to be vastly more efficient in the use of uranium while 'burning' radioactive substances that otherwise would have to be disposed of as waste. BREST refers to *bystry reaktor so svintsovym teplonositelem*, Russian for 'fast reactor with lead coolant'. Its core would measure about 2.3 metres in diameter by 1.1 metres in height and contain 16 tonnes of fuel. The unit would be refuelled every year, with each fuel element spending five years in total within the core. Lead coolant temperature would be around 540 °C, giving a high efficiency of 43%, primary heat production of 700 MWt yielding electrical power of 300 MWe. The operational lifespan of the unit could be 60 years. The design is expected to be completed by NIKIET in 2014 for construction between 2016 and 2020.

Construction of the BN-800 reactor

On February 16, 2006, the U.S., France and Japan signed an "arrangement" to research and develop sodium-cooled fast reactors in support of the Global Nuclear Energy Partnership. In April 2007 the Japanese government selected Mitsubishi Heavy Industries as the "core company in FBR development in Japan". Shortly thereafter, MHI started a new company, Mitsubishi FBR Systems (MFBR) to develop and eventually sell FBR technology.

The Marcoule Nuclear Site in France, location of the Phénix (on the left) and possible future site of the ASTRID Gen-IV reactor.

In September 2010 the French government allocated 651.6 million euros to the *Commissariat à l'énergie atomique* to finalize the design of "Astrid" (Advanced Sodium Technological Reactor for Industrial Demonstration), a 600 MW reactor design of the 4th generation to be operational in 2020. As of 2013 the UK had shown interest in the PRISM reactor and was working in concert with France to develop ASTRID.

In October 2010 GE Hitachi Nuclear Energy signed a memorandum of understanding with the operators of the US Department of Energy's Savannah River Site, which should allow the construction of a demonstration plant based on the company's S-PRISM fast breeder reactor prior to the design receiving full NRC licensing approval. In October 2011 The Independent reported that the UK Nuclear Decommissioning Authority (NDA) and senior advisers within the Department for Energy and Climate Change (DECC) had asked for technical and financial details of the PRISM, partly as a means of reducing the country's plutonium stockpile.

The traveling wave reactor proposed in a patent by Intellectual Ventures is a fast breeder reactor designed to not need fuel reprocessing during the decades-long lifetime of the reactor. The breed-burn wave in the TWR design does not move from one end of the reactor to the other but gradually from the inside out. Moreover, as the fuel's composition changes through nuclear transmutation, fuel rods are continually reshuffled within the core to optimize the neutron flux and fuel usage at any given point in time. Thus, instead of letting the wave propagate through the fuel, the fuel itself is moved through a largely stationary burn wave. This is contrary to many media reports, which have popularized the concept as a candle-like reactor with a burn region that moves down a stick of fuel. By replacing a static core configuration with an actively managed "standing wave" or "soliton" core, TerraPower's design avoids the problem of cooling a highly variable burn region. Under this scenario, the reconfiguration of fuel rods is accomplished remotely by robotic devices; the containment vessel remains closed during the procedure, and there is no associated downtime.

Breeding

Breeding in nuclear reactors refers to the process in which significantly amount of fertile materials are converted to fissile materials by nuclear transmutation. This requires the fertile isotope to have large cross section for neutron capture. Since the main purpose of a nuclear reactor is to produce electricity, breeding is considered as an off-shoot of excess neutrons produced during fission above the ones required for sustenance of chain reaction.

The possibility of breeding in a nuclear reactor, taking into account of the type of fissile material used, depends on the number of neutrons produced for every neutron absorbed in the fuel. This is denoted by reproduction factor 'η'. This is related to the cross sections as follows:

$$\eta = \frac{\sigma_f}{\sigma_a}\nu \qquad (9)$$

The probability of breeding is enhanced when the value of 'η' exceeds two by a large fraction. For example, the value of 'η' for U-235, Pu-239 and U-233 when bombarded by thermal neutrons is 2.07, 2.11 and 2.30 respectively. Breeding with thermal neutrons using U-235 and Pu-239 fuels is virtually impossible due to neutron absorption in structural materials and moderator. However, with U-233 fuel, it is possible to achieve breeding using thermal neutrons.

The scenario is different for the case of bombardment with fast neutron. The value of 'η' for U-235, Pu-239 and U-233 when bombarded by fast neutrons is 2.3, 2.7 and 2.45 respectively. Hence, breeding is possible with all the above fissile nuclei due to bombardment with fast neutron.

Conversion ratio is a widely used term to denote the ability of a reactor to convert fertile to fissile material. Conversion ratio is defined as the ratio of number of fissile nuclei produced to the number of fissile nuclei consumed.

When a reactor is operated without any fertile material, the conversion ratio is zero. In this case, the reactor is referred to as 'burner' as it burns all the fuel without producing any fissile material.

It may be recalled that the uranium fuelled thermal reactors either use natural uranium or enriched uranium. In both these cases, isotopic abundance of U-238, a fertile material, is very high. It is possible to achieve a higher conversion ratio by facilitating higher neutron capture in U-238 relative to neutron absorption in U-235. This must be carried out without the loss of criticality. To ensure that sufficient neutrons are available for chain reaction, the neutron losses in structural elements and moderator must be reduced. Also, the use of moderators like heavy water and carbon require more collisions for neutron thermalization and hence require larger core. This improves the contact between neutron and U-238 contact, thereby improving the conversion ratio.

The conversion ratio can be predicted approximately as follows:

$$Conversion\ Ratio\,(CR) = \eta_{235^{\varepsilon-1-l}} \qquad (10)$$

'η_{235}' corresponds to reproduction factor of pure U-235. Hence to improve the conversion ratio, leakage of neutrons (l) must be reduced. For most thermal reactors, CR is between 0.4 and 0.7 and these reactors are called 'converters' (0<CR<1). 'ε' is the fast fission factor that indicates the contribution due to fast neutrons in a thermal reactor.

Breeding Ratio

The reactor with conversion ratio greater than 1 is called a breeder reactor. This is the reactor that produces more fuel than that it consumes. For breeder reactors, the term 'breeding ratio (BR)' is used.

The breeding ratio is maximum when the leakage of neutrons (l) is zero. This is called maximum breeding ratio (BR_{max}) and is also called Breeding Potential of the fuel.

$$Breeding\ \ Potential = BR_{max} = \eta - 1 \qquad (11)$$

Please note that in Eq. (11), 'ε' the fast fission factor is taken as unity as most of the breeders are fast breeders. The reproduction factor is to be calculated for Pu-239, the predominant fissile isotope in fast reactors.

Breeding Gain

Another term widely used in a breeder reactor is 'Breeding Gain (BG)'. The relationship between BG and BR is

$$BG = BR - 1 \quad (12)$$

This represents the extra fissile material produced for every atom of fuel (fissile isotope) consumed.

For a Pu-239 fuelled fast reactor,

$$BR = \eta_{239} - 1 - l \quad (13)$$

$$CR = BR = \left(\frac{\sigma_f}{\sigma_a} \nu \right)_{239} - 1 - l \quad (14)$$

Example -1: Determine Breeding Ratio for Pu-239 fuelled fast reactor. Take $\nu = 2.975$; $\sigma_f = 1.850$; $\sigma_a = 2.11$ and $l = 0.405$.

Using Eq. (5), $BR = \left(\frac{1.85}{2.11} 2.975 \right) - 1 - 0.405$

BR = 1.203

Example – 2: Consider a fast breeder reactor operating with Breeding Ratio of 1.3. If it is desired to accumulate an additional 1500 kg of fissile material, determine the amount of pure Plutonium fuel to be burnt.

By definition of Breeding Ratio,

BR = number of fissile nuclei produced/number of fissile nuclei consumed

BR = mass of fissile nuclei produced/mass of fissile nuclei consumed

Let 'x' be the mass of fissile nuclei consumed, then BR = (x+1500)/x=1.3 Solving the above for 'x' gives x = 5000 kg

Hence 5000 kg of Plutonium must be burnt to produce an additional 1500 kg of fissile material.

Doubling Time (DT)

It is defined as the time required to accumulate a mass of fuel equal to that loaded initially in a reactor system. When the initial inventory of fissile material is low, doubling time is reduced. In other words, doubling is achieved in a shorter period of time with initial lower loading of fissile material.

Let us derive an expression for calculation of Doubling Time (DT).

The reactor power per unit mass of fuel (P') may be related to the number of fuel nuclei per unit mass of the fuel (N_f) as

$$P' = E_f N_f \bar{\phi} \sigma_f \quad (15)$$

The above equation may be rewritten as:

Power per unit mass of fuel (W/kg)= Energy released per fission (J) * Number fissions per unit time per unit mass of fuel ($s^{-1}kg^{-1}$)

The number of fissions per unit time per unit mass of fuel represents the rate of consumption of a unit mass of fuel.

Therefore, If M_F is the mass of fuel loaded, then the rate of consumption (R_C) of fuel of mass M_F is given by

$$R_C = M_F N_f \phi^- \sigma_a \text{ (atoms/s)}$$

If 'BR' is the Breeding Ratio, then the rate of production (R_p) of fissile mass is given by

$$R_P = BR*M_F N_f \phi^- \sigma_a \text{ (atoms/s)}$$

The net increase in fuel = Rate of Production – Rate of consumption

$$\text{Net increase} = R_{net} = (BR-1)*M_F N_f \phi^- \sigma_a \text{ (atoms/s)}$$

If DT is the Doubling Time, then

$$DT*R_{net} = M_F N_f$$

Therefore,

$$DT = \frac{M_F N_f}{R_{net}} = \frac{M_F N_f}{(BR-1)M_F N_f \phi^- \sigma_a} = \frac{1}{(BR-1)\phi^- \sigma_a}$$

An expression may be obtained for DT In terms of reactor power (P') as follows:

$$\text{Recall, } P' = E_f N_f \phi^- \sigma_f$$

$$\phi^- = P' / (N_f E_f \sigma_f)$$

Substituting the above in the equation for Doubling Time, we get

$$DT = \frac{N_f E_f \sigma_f}{(BR-1)\sigma_a P'}$$

Now let's discuss the factors that influence Doubling Time.

(i) Breeding Ratio: Higher the Breeding Ratio, greater is the amount of fissile material produced. Hence, the time required for doubling the mass is reduced at higher Breeding Ratios. We have seen earlier that Breeding Ratio depends on the reproduction factor h, which inturn is influenced by absorption cross section, fission cross sections and 209½'. As the cross sections are functions of neutron energies, appropriate choice of neutron energy and steps to minimize neutron leakage will result in increased Breeding Ratio (BR).

Power per unit mass of the fuel: When this quantity is high, the average neutron flux is also high among other variables. As evident from Eq. (11), with increased neutron flux, the Doubling Time is reduced. Hence the Doubling time decreases with increased reactor Power per unit mass of the fuel.

Example – 4: Determine the doubling time of a Pu-239 fast breeder reactor. The reactor is operated at 400 MW/tonne Pu-239 with a Breeding Ratio of 1.2. The absorption and fission cross section are 2.16 and 1.81 b respectively. The number of Pu-239 per unit mass is $2.52x10^{21}$ (atoms/g).

Data: $E_f = 3.2x10^{-11}$J; $N_f = 2.52x10^{21}$; P' = 400 MW/tonne = 400 J/gs;

$\sigma_a = 2.16$ b; $\sigma_f = 1.81$ b

Substituting the above in Eq. (13),

$$DT = \frac{N_f E_f \sigma_f}{(BR-1)\sigma_a P'} = \frac{2.52x10^{21} * 3.2x10^{-11} * 1.81}{(1.2-1)2.16 * 400}$$

DT = 9776 days 27 years

Need for Breeding

Per-capita energy consumption of a nation is an indication of growth and prosperity of a nation. As all of us are aware that energy consumption throughout the world is increasing continuously due to several reasons that include population increase, better standard of living in terms of comfort and industrialization. Hence any mode of energy generation meant to serve industry and society needs to be sustainable. The problem of rapid depletion of oil and coal reserves is well known. One method of achieving sustainability is to improve the efficiency of energy generation/conversion such that the energy generated from unit mass of fuel is maximized. Another method is the re-use of material in one form or the other in order to minimize the amount of fresh fuel inventory required. Hence inline with these principles, power generation from nuclear fuel must also be made sustainable for long-term use.

The natural uranium contain very little amount of fissile isotope (0.7 %) while the rest is U-238. Hence in typical pressurized heavy nuclear reactors, to utilize one kg of fissile isotope for power generation, nearly 100 kg of natural uranium is used. During fission reaction by neutron bombardment, the fissile material is destroyed or consumed. If fissile material could be produced from fertile material available abundantly in the core, inventory of fresh fuel required can be reduced. The conversion of fertile material to fissile material due to neutron irradiation in the core can be achieved by one of the nuclear reactions called nuclear transmutation. This is called breeding in which fissile isotopes are produced from fertile isotopes.

Nuclear reactors that produce more fissile isotope than they consume are called breeders or breeder reactors. Depending upon the neutron energy utilized for breeding, breeders are classified as fast breeders or thermal breeders.

Comparison of Thermal and Fast Spectrum for Breeding

Fast Breeders or Fast Breeder Reactors (FBR) are the common breeder reactors utilized in the world. For successful breeding, at least one neutron must be exclusively available for interaction with fertile isotope for every neutron absorbed by fuel, after taking into account of neutron absorption in core and structural materials. For example, for every fast neutron absorbed by Pu-239, 2.45 neutrons are produced. If one neutron is utilized for sustenance of chain reaction and on an average 0.45 neutron is lost, the remaining one neutron is available for conversion of fertile to fissile isotope. With U-235, an average of 2.1 neutrons are produced for every neutron absorbed. After accounting for neutrons required for sustenance of chain reaction and for losses, on an average less than one neutron is available which is insufficient for breeding. In other words, the number of neutrons produced per neutron absorbed must be greater than 2 by atleast a certain amount to facilitate breeding.

Now let us analyze the scenario in thermal spectrum. The number of neutrons produced per neutron absorbed in Pu-239 and U-235 due to bombardment of thermal neutrons is 2.04 and 2.06 respectively, clearly indicating that thermal breeders with these fissile isotopes are impractical. Hence most of the breeder reactors are fast reactors utilizing Pu-239 as the fissile isotope. Hence such breeder reactors are called fast breeder reactors. Fast breeder technology for power generation is confined only to few countries like France, Russia, China, India, Japan and South Korea. It may be noted that thermal reactors are used in several countries for power generation while only the above countries have active fast breeder programme.

One of the less-explored or lesser known fissile material is U-233. This isotope does not occur in nature and hence produced artificially in reactors using Th-232 as fertile isotope. In Indian fast reactor, Th-232 is used in the blanket, from which U-233 is produced. A fast reactor operating with U-233 as the fissile material will produce 2.31 neutrons per every neutron absorbed and hence with appropriate core design aimed at minimizing neutron loss, a fast breeder reactor operating with U-233 as fissile material can also be built.

Breeding is possible even with thermal neutrons if U-233 is used as fissile material, as 2.26 neutrons are produced per neutron absorbed. A breeder reactor utilizing thermal neutrons is called thermal breeder. India is developing an Advanced Heavy Water Reactor that will act as thermal breeder utilizing U-233 as fissile isotope and Th-232 as fertile isotope. Availability of large reserves of thorium in India (world's 1/3 of total thorium reserves) is a prime reason for India's interest in breeders utilizing U- 233.

Table: Number of neutrons produced per neutron absorbed for different isotopes in fast and thermal spectrum

Neutron Energy	Number of neutrons produced per neutron absorbed (η)		
	U-235	Pu-239	U-233
0.025 eV	2.06	2.04	2.26
> 1 MeV	2.10	2.45	2.31

It is to be noted that the number of neutrons produced per neutron absorbed (h), also called 211reproduction factor shown in Table is not a measured quantity. It is determined from other measurable quantities as follows:

$$\eta_0=\upsilon\sigma_f/\sigma_a = \upsilon\sigma_f/(\sigma_c+\sigma_a) = \upsilon/(1+\alpha)$$

In the above equation, σ_a, σ_f and σ_c refer to absorption, fission and capture cross sections. 'υ' refers to the number of neutrons produced per fission. 'η_0' is the number of neutrons produced per neutron absorbed for a pure fissile material. When fuel is a mixture of fertile and fissile material, reproduction factor (η) is

$$\eta=(N_1\upsilon_1\sigma_{f1}+N_2\upsilon_2\sigma_{f2})/(N_1\sigma_{a1}+ N_2\sigma_{a2})$$

In Equation above, the suffices 1 and 2 represent the isotopes 1 and 2 (say U-235 and U- 238); N_1 and N_2 are the number fractions of isotopes 1 and 2.

The following example shows calculation of reproduction factor for a typical scenario.

Example – 1: Determine the reproduction factor in a fast reactor operating with a pure fuel whose fission and absorption cross sections are 1.39 b and 1.64 b respectively. The average number of neutrons produced per fission is 2.605

Solution:

Recall earlier Eq.

$$\eta_0=\upsilon\sigma_f/\sigma_a = \upsilon\sigma_f/(\sigma_c+\sigma_a) = \upsilon/(1+\alpha)$$

$\sigma_f = 1.39$ b

$\sigma_a = 1.64$ b

$\upsilon = 2.605$

Therefore, reproduction factor (η_0) = 2.207

Example – 2: The following data pertain to a nuclear reactor utilizing natural uranium (N_{U-235}/N_{U-238}=0.007) and operating in thermal spectrum:

U-235	U-238
σ_f = 586 b	σ_f = 0 b
σ_a = 681 b	σ_a = 2.7 b
υ = 2.42	υ = 0

Determine the reproduction factor.

Solution:

Let '1' and '2' represent the isotopes U-235 and U-238 respectively. Using earlier Eq.,

$$\eta=(N_1\upsilon_1\sigma_{f1}+N_2\upsilon_2\sigma_{f2})/(N_1\sigma_{a1}+ N_2\sigma_{a2})$$

Dividing the numerator and denominator of right side of the above Equation by N2, we get

$$\eta=(N_1\upsilon_1\sigma_{f1}/N_2+\upsilon_2\sigma_{f2})/(N_1\sigma_{a1}/N_2+\sigma_{a2})$$

$$\eta=(0.007*2.42*586)/(0.007*681+2.7)=1.33$$

Therefore, the reproduction factor under the above circumstances is 1.33

Example – 3: The following data correspond to U-233 in a thermal reactor.

$\sigma_f = 531$ b

$\sigma_a = 576$ b

$\upsilon = 2.49$

Determine the reproduction factor if the fuel is pure.

Solution:

Recall Eq. (1),

$$\eta_0=\upsilon\sigma_f/\sigma_a$$

Substituting the data in Eq. (1), reproduction factor is obtained as 2.30

Example - 4: If a fast reactor is loaded with U-235 and U-238 such that $N_{U\text{-}235}/N_{U\text{-}238}=0.25$, comment on the possibility of breeding. Data on cross section and average number of neutrons produced per fission are given below:

U-235	U-238
$\sigma_f = 1.4$ b	$\sigma_f = 0.095$ b
$\sigma_a = 1.65$ b	$\sigma_a = 0.255$ b
$\upsilon = 2.60$	$\upsilon = 2.60$

Solution:

Let '1' and '2' represent the isotopes U-235 and U-238 respectively. Using Eq. (2),

$$\eta=(N_1\upsilon_1\sigma_{f1}+N_2\upsilon_2\sigma_{f2})/(N_1\sigma_{a1}+N_2\sigma_{a2})$$

$$\eta=(N_1\upsilon_1\sigma_{f1}/N_2+\upsilon_2\sigma_{f2})/(N_1\sigma_{a1}/N_2+\sigma_{a2})$$

$$\eta=(0.25*2.6*1.4+2.6*0.095)/(0.25*1.65+0.255)=1.73$$

The reproduction factor is 1.73 and hence it is not possible to achieve breeding with this configuration.

Hence one may understand from Examples – 1 to 5, that the reproduction factor could be varied by using fuels with different enrichment levels or composition and by use of neutrons of appropriate energy (fast neutron or thermal neutron). In summary, fast breeder reactors are practical with Pu-239 as the fissile material, while thermal breeders can be built with U-233 as fuel.

Nuclear Fuel Cycle

We will look at the 'Nuclear Fuel Cycle' to understand the way the fuel is handled from its mining till its disposal. The fuel cycle comprises the processes carried out to transform ore to the fuel suitable for use in core (called front end of the cycle), transformations to the fuel while it is being used in the reactor (called service period) and the activities performed to deal with the spent fuel from the reactor (called back end of the cycle).

There are two types of fuel cycles: Open fuel cycle and Closed fuel cycle.

Open Fuel Cycle

The cycle begins with the extraction of uranium ore from mine followed by processing the extracted uranium. Enrichment is carried out to increase the percentage of U-235 from 0.7 % (natural uranium) to 2-3.5 % required for light water reactors.

For use in heavy water reactors, enrichment is not required. Fuel is fabricated in the desired form and loaded in the core. These constitute front end of the cycle.

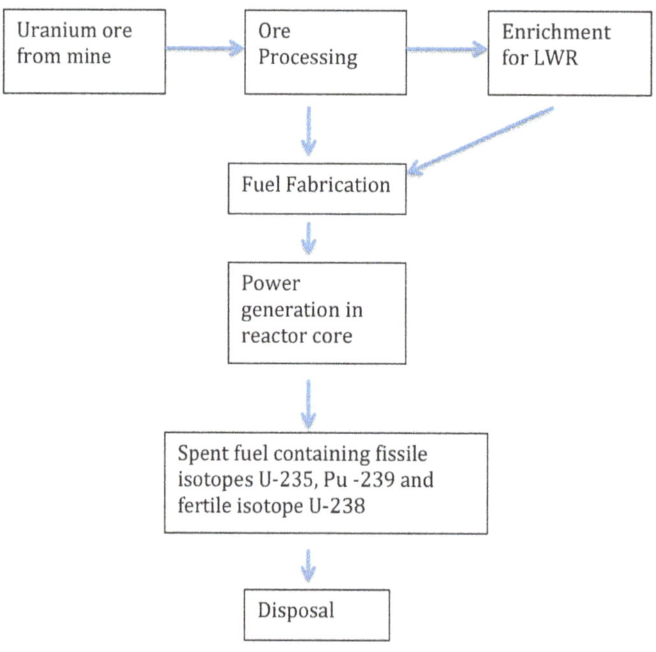

Schematic layout of open fuel cycle

After the fuel has been irradiated for power generation (service life), the fuel from core is removed as spent fuel. The spent fuel too contains fissile isotopes like U-235 and Pu-239 but in lower amounts that do not sustain chain reaction. Apart from fissile isotopes, fertile isotopes, minor actinides and activation products are also present in spent fuel. Safe disposal of spent fuel completes the cycle wherein no effort is made to separate the components of spent fuel.

Note: Activation products are the products formed due to the neutron absorption by non-fuel components of reactor like coolant, control rod and structural materials.

Closed Fuel Cycle

The closed fuel cycle is similar to that of open fuel cycle during front end of the cycle and service periods. The difference lies in the handling of spent fuel. While no attempt is made to reprocess the spent fuel in open cycle, the same is reprocessed in closed fuel cycle. The reprocessing is aimed at separating the plutonium (produced by nuclear transmutation) and unutilized uranium from other components of spent fuel. The recovered uranium and plutonium are recycled for use as fuel in nuclear reactors. Hence through this cycle, better utilization of resources can be made. The products of fission are separated and then sorted depending on their half lives and activities, before disposing them appropriately without excessive burden to environment.

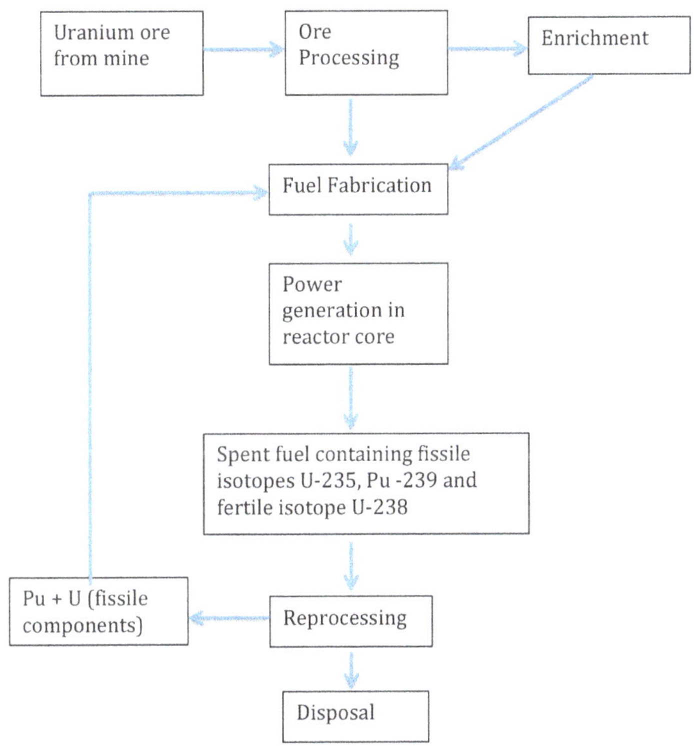

Schematic layout of closed fuel cycle

Most light water reactors fo with the laws prevailing in the respective country. Appropriate precautions are taken to store the spent fuel safely in an isolated region.

With the known conventional resources of uranium and with the number of light water reactors in operation and those planned for future, uranium is expected to be available for about 80 years only. With additional uranium resources still unexplored this may extend to about 270 years.

Fast breeder reactors use closed fuel cycle. A typical core of fast breeder reactor contains a mixture of UO_2 and PuO_2. While U-235 in UO_2 and Pu-239 in PuO_2 act as fissile material generating energy, the abundant U-238 in blanket undergoes nuclear transmutation producing more Pu-239. As a result the energy that can be produced from uranium in a fast breeder reactor is significantly more than that produced in a thermal reactor. On a conservative estimate, energy potential in fast breeder reactor is about 60 times greater than that in a thermal reactor for the same initial uranium inventory. If all the existing thermal reactors were replaced by fast breeder reactors, it would be possible to utilize the known conventional resources of uranium to operate these reactors for about 4500-5000 years (80*60), a time period high enough to consider the source to be inexhaustible.

The higher energy potential in fast breeder reactor means that the power cost can be independent of fluctuations in fuel cost. The fresh fuel to be replenished is minimal, if not zero and hence makes it possible to de-link cost of power production from the fluctuations in the price of fuel.

With the closed fuel cycle possible with fast reactors, the quantity of accumulated long-lived isotopes that must be disposed is likely to be less than 0.1 % of the fission products. This reduces the burden of waste management by reducing the storage space required for disposal. Also the time for which the fuel needs to be stored to bring the toxicity down to natural levels is reduced considerably, say by 1000 times. In nutshell, with closed fuel cycle adopted in Fast Breeder Reactor, the following advantages are envisaged:

(i) reduction in mining of fresh uranium

(ii) considerable reduction in waste management time and money

In summary, with the use of fast breeder reactor nuclear energy may be made an inexhaustible source due to increased utilization of energy potential of uranium, independence on fluctuations in fuel cost and lesser waste generation.

References

- U. Mertyurek; M. W. Francis; I. C. Gauld. "SCALE 5 Analysis of BWR Spent Nuclear Fuel Isotopic Compositions for Safety Studies" (PDF). ORNL/TM-2010/286. OAK RIDGE NATIONAL LABORATORY. Retrieved 25 December 2012
- Waltar, A.E.; Reynolds, A.B (1981). Fast breeder reactors. New York: Pergamon Press. p. 853. ISBN 978-0-08-025983-3
- Rodriguez, Placid; Lee, S. M. "Who is afraid of breeders?". Indira Gandhi Centre for Atomic Research, Kalpakkam 603 102, India. Retrieved 24 December 2012
- Helmreich, J.E. Gathering Rare Ores: The Diplomacy of Uranium Acquisition, 1943–1954, Princeton UP, 1986: ch. 10 ISBN 0-7837-9349-9
- Milsted, J.; Friedman, A. M.; Stevens, C. M. (1965). "The alpha half-life of berkelium-247; a new long-lived isomer of berkelium-248". Nuclear Physics. 71 (2): 299. doi:10.1016/0029-5582(65)90719-4
- R. Bari; et al. (2009). "Proliferation Risk Reduction Study ofAlternative Spent Fuel Processing" (PDF). BNL-90264-2009-CP. Brookhaven National Laboratory. Retrieved 16 December 2012
- Frank von Hippel; et al. (February 2010). Fast Breeder Reactor Programs: History and Status (PDF). International Panel on Fissile Materials. ISBN 978-0-9819275-6-5. Retrieved 28 April 2014
- C.G. Bathke; et al. (2008). "An Assessment of the Proliferation Resistance of Materials in Advanced Fuel Cycles" (PDF). Department of Energy. Retrieved 16 December 2012

Permissions

All chapters in this book are published with permission under the Creative Commons Attribution Share Alike License or equivalent. Every chapter published in this book has been scrutinized by our experts. Their significance has been extensively debated. The topics covered herein carry significant information for a comprehensive understanding. They may even be implemented as practical applications or may be referred to as a beginning point for further studies.

We would like to thank the editorial team for lending their expertise to make the book truly unique. They have played a crucial role in the development of this book. Without their invaluable contributions this book wouldn't have been possible. They have made vital efforts to compile up to date information on the varied aspects of this subject to make this book a valuable addition to the collection of many professionals and students.

This book was conceptualized with the vision of imparting up-to-date and integrated information in this field. To ensure the same, a matchless editorial board was set up. Every individual on the board went through rigorous rounds of assessment to prove their worth. After which they invested a large part of their time researching and compiling the most relevant data for our readers.

The editorial board has been involved in producing this book since its inception. They have spent rigorous hours researching and exploring the diverse topics which have resulted in the successful publishing of this book. They have passed on their knowledge of decades through this book. To expedite this challenging task, the publisher supported the team at every step. A small team of assistant editors was also appointed to further simplify the editing procedure and attain best results for the readers.

Apart from the editorial board, the designing team has also invested a significant amount of their time in understanding the subject and creating the most relevant covers. They scrutinized every image to scout for the most suitable representation of the subject and create an appropriate cover for the book.

The publishing team has been an ardent support to the editorial, designing and production team. Their endless efforts to recruit the best for this project, has resulted in the accomplishment of this book. They are a veteran in the field of academics and their pool of knowledge is as vast as their experience in printing. Their expertise and guidance has proved useful at every step. Their uncompromising quality standards have made this book an exceptional effort. Their encouragement from time to time has been an inspiration for everyone.

The publisher and the editorial board hope that this book will prove to be a valuable piece of knowledge for students, practitioners and scholars across the globe.

Index

www.ingramcontent.com/pod-product-compliance
Lightning Source LLC
Chambersburg PA
CBHW082043190326
41458CB00010B/3447